THE COMET BOOK

THE COMET BOOK

A GUIDE FOR THE RETURN OF HALLEY'S COMET

ROBERT D. CHAPMAN & JOHN C. BRANDT

Laboratory for Astronomy and Solar Physics
NASA/Goddard Space Flight Center
Greenbelt, Maryland

Jones and Bartlett Publishers, Inc.
Boston   *Portola Valley*

Editorial offices: Jones and Bartlett Publishers, Inc., 30 Granada Court, Portola Valley, CA 94025.

Sales and customer service offices: Jones and Bartlett Publishers, Inc., 20 Park Plaza, Boston, MA 02116.

Cover illustration: Comet Bradfield observed at approximately 17:00 G.m.t. on January 10, 1980. The image was made using a low-resolution imaging system on the International ultraviolet Explorer that is used to point the spacecraft at targets for spectroscopic observations. The system produces black and white images of celestial objects, but the images are usually displayed in false color to show clearly the large brightness range that can be recorded. In this image, light blue represents the lowest intensity level and deep red the highest. The intensity scale at the bottom illustrates the intermediate levels. The image has been processed at the Interactive Astronomical Data Analysis Facility of the Laboratory for Astronomy and Solar Physics at the Goddard Space Flight Center to remove plus-shaped fiducial marks introduced by the cameras. (Image obtained by Michael A. A'Hearn of the University of Maryland and reprocessed by D. A. Klinglesmith, III, of the Goddard Space Flight Center.)

Indication of the authors' affiliation with the National Aeronautics and Space Administration on the title page of this book does not constitute endorsement of any of the material contained herein by NASA. The authors are solely responsible for any views expressed.

Library of Congress Cataloging in Publication Data

Chapman, Robert Dewitt
    The comet book.

    Bibliography: p.
    Includes index.
    1. Halley's comet. I. Brandt, John C. II. Title.
QB723.H2C46   1984   523.6'4   84-12569
ISBN 0-86720-029-4

STAFF FOR THIS BOOK: *Publisher:* Arthur C. Bartlett.
Designer: Elizabeth W. Thomson   *Illustrator:* John Hamwey.
Production: Unicorn Production Services, Inc.
Composition: Optima, set by Achorn Graphic Services, Inc.
Printing and binding: Halliday Lithograph Corp.
Color printing: New England Book Components

Printed in the United States of America.   10   9   8   7   6   5   4   3   2

PREFACE

Late in 1982, astronomers using electronic cameras at the cutting edge of technology together with the great 5-meter Hale Telescope on Palomar Mountain in California found Comet Halley as it once again moved in toward the sun. As this comet continues on its way to its 1985–1986 encounter with the earth, scientists all over the world are in high gear preparing for a flurry of research activity that is without precedent in the history of cometary science. As Comet Halley nears the earth, increasing excitement will be generated by public interest in comets in general and Comet Halley in particular. We have written this book to help satisfy people's curiosity about these fascinating celestial objects. The narrative has a twofold purpose: to present a readable and scientifically sound description of comets, and to show how comets have influenced our view of the universe over the ages.

The book begins with a description of the arrival of a bright comet in the night sky and a brief discussion of its appearance and physical properties. It then goes on to a brief history of comet studies; a comprehensive, but elementary, description of the origin of comets and the physical processes occurring in them; a summary of the latest plans for the study of Comet Halley, including missions to be launched by the European Space Agency, the Soviet Union, and Japan; and a description of the United States' mission to Comet Giacobini–Zinner, the first direct mission to a comet.

The 1985–1986 appearance of Comet Halley is discussed in some detail. There is extensive material to help the reader observe the comet, including charts of its changing position on the sky for observers in both the northern and southern hemispheres. Unique perspective diagrams show the orbit of Comet Halley for each appearance since 1759. For inexperienced observers, there are discussions of how to observe the comet with binoculars and to photograph it with a single-lens reflex camera.

The book is profusely illustrated with line drawings and photographs, including some of the best color photographs ever made of comets. In addition to photographs of many important comets from the past, there are numerous pictures of Comet Halley at previous passes and the 1982 rediscovery photograph. Another feature of the book is a collection of pictures of some of the most important historical and modern comet scientists. An extensive glossary defines unfamiliar terms used in the text.

As professionals in the field of comet science, we have devoted a great deal of our professional time to the study of comets. We are extremely excited about these objects and especially about the impending return of Comet Halley. We hope this book will enable you to share this excitement.

Robert D. Chapman
John C. Brandt

CONTENTS

CHAPTER 1

A COMET APPEARS

OBSERVING A COMET

It is a crisp, clear, early March morning. The thermometer on my porch stands at a few degrees below freezing. The constellations are arrayed across the sky as they would be early on a summer evening. Leo the lion is diving nose first toward the western horizon, and Scorpius is rising in the southeast. Soon morning twilight will begin to tint the eastern horizon. Comet West stands out crisply against a still dark northeastern sky, a few degrees above the horizon.

The bright head of the comet lies near the horizon in the constellation Pegasus, the winged horse. The comet's tail sweeps upward, reaching nearly 20° toward Cygnus, the swan. With my unaided eye I can make out two components in the tail: a narrow, very straight ray, nearly perpendicular to the horizon, and a broad, more diffuse tail, curving off to my left (Figure 1–1 and Plate 1). The sight is so enchanting that I forget the cold, and the moment of displeasure I felt when the alarm clock sounded at 4:00 A.M.

Dawn is approaching. A red glow creeps up from the eastern horizon, obliterating the stars one by one. The brightening sky soon overwhelms the comet. I feel some regret, and anticipation too, as I head off to the day's business. There will be more mornings to view Comet West as it moves away from the sun. Its fading brightness will be partially offset by its increasing distance from the horizon, so it can be observed longer and longer before sunrise. Even so, it should fade from naked eye view in a few weeks.

Scenes like this are repeated with surprising frequency. History shows that at least one comet reaches naked eye brightness almost every year. Comets are, therefore, one type of celestial object that anyone can observe easily. These fascinating objects are brightest when their highly elongated paths through space bring them near the sun. It should be no surprise, then, that the most conspicuous comets,

Figure 1–1. Comet West as photographed from near White Sands, New Mexico, on the morning of March 4, 1976. (Courtesy of A. Stober, Laboratory for Astronomy and Solar Physics, NASA/Goddard Space Flight Center, Greenbelt, Maryland.)

like Comet West, tend to be observed in the direction of the sun on the sky: the eastern sky just before sunrise or the western sky just after sunset.

    The main components of a bright comet are visible to the naked eye or to an observer using binoculars (Figure 1–2). The head is the brightest part. Sometimes a star-like point of light, called the nucleus, is visible inside the head. Careful study of the head reveals that its brightness gently decreases away from the nucleus. It is not always obvious from observations whether what appears to be the nucleus is a separate entity inside the head, or just the brightest part of the head. Sometimes astronomers refer to a false nucleus, which is merely the brightest part.

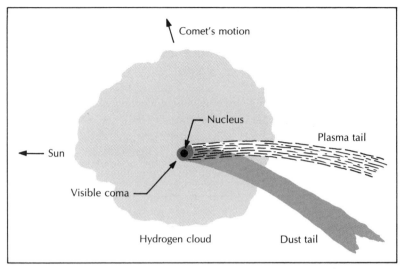

Figure 1–2. The principal parts of a comet (not drawn to scale).

The remainder of the head surrounding the nucleus is called the coma. Figure 1–2 also shows the huge hydrogen cloud around the nucleus (see Chapter 6) that is *not* visible to the eye. It was discovered by the use of spacecraft carrying instruments sensitive to ultraviolet light.

Extending from the coma is the tail. Like Comet West, many comets have composite tails. Astronomers have arbitrarily called the narrow, straight tails "type I" tails and the broad, curving tails "type II" tails. There are other names for these tails—plasma tails and dust tails—based on an understanding of their nature. Astronomers often assign terms that are quite arbitrary to newly discovered phenomena. This is necessary for communication among workers in the field. As understanding progresses, the arbitrary terms are replaced by terms that contain some information about the physical nature of the phenomena. Such names tend to be easier to remember and are much more meaningful. For the moment, the terms "plasma tail" and "dust tail" may be as mysterious as "type I" and "type II," but the reason the former are more meaningful will soon appear.

With binoculars or a small telescope, it is possible to observe an interesting phenomenon inside the comas of some comets. Narrow curved structures resembling spirals or streams of water in a fountain can be seen. These structures are sometimes seen to change shape over a period of several hours, if the comet is visible for sufficiently long. Bright knots of material or kinks in a comet's tail are also observed. These too may change during several hours' time.

You too can become a comet observer! To find a comet, you need to consult a popular magazine such as *Sky and Telescope* or *Astronomy*, or a newspaper that announces the positions of bright comets, whenever one comes around. If you observe on many successive nights, you can see the tail grow and shrink as the comet approaches or recedes from the sun. You can compare the brightness of the head with bright stars and see its brightness change from night to night. If you use binoculars or a low-power telescope, you might even observe the nucleus and see jets or knots moving in the comet. Give it a try! You are likely to find it fun, and intellectually rewarding.

WHAT IS A COMET?

The main component of a comet is its nucleus, a chunk of ice with lots of cosmic dust imbedded in it. A typical nucleus is really not very large, perhaps only a few kilometers in diameter. The nucleus travels around the sun in an elliptical orbit, much as the planets do. However, while planetary orbits are nearly circular, cometary orbits are typically very elongated. As the nucleus travels around its orbit, it moves at a speed that depends on its distance from the sun. Generally, the nucleus moves very slowly when it is in the cold, outer reaches of the solar system, far from the sun. However, when the nucleus is near the sun it moves very fast. Thus a typical comet spends most of its time far from the sun and the earth and is not very conspicuous during that time.

When the iceberg nucleus approaches the sun, things begin to happen that give rise to all the fascinating phenomena we observe. Solar radiation begins to heat the surface of the nucleus. The ice sublimes; that is, it changes from the solid state to the gaseous state without becoming a liquid. In the process, small grains of material (which we call dust) that were imbedded in the ice are released into space. The gas and dust grains form a large halo around the nucleus; this is the coma. Usually the coma grows as a comet moves in toward the sun, reaching a maximum size when the comet is about twice as far from the sun as is the earth. At that time, a coma can be 100,000 kilometers across or even larger.

Streaming from the coma, always pointing away from the sun (Figure 1–3), is the comet's tail. Well before the invention of the telescope, students of nature had noticed that comets are always positioned with their heads closest to the sun and their tails pointing away from the sun, but it was not until the early twentieth century that the physical significance of this observation began to be understood. The

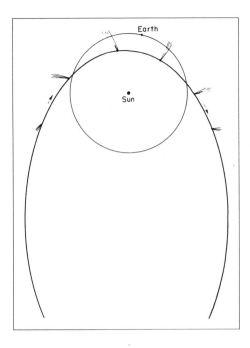

Figure 1–3. Comet tails point away from the sun, as shown in this drawing made by E. E. Barnard in the late nineteenth century (Yerkes Observatory photograph).

tail, like the coma, grows as the comet approaches the sun, reaching typical lengths of 10,000,000 kilometers—and in rare cases, several times 100,000,000 kilometers. There is a lot of fascinating physics going on in comet tails; we will look at some details in Chapter 6.

To help understand later discussions, let us compare the motions of comets and planets. The planetary system is flat. The orbit of each planet lies in a plane, and the orbit planes are inclined to one another by small angles. The plane of the earth's orbit is frequently taken as the reference plane. If we extend this plane until it intersects the sky, it defines a circle known as the ecliptic. As seen from the earth, the sun, moon, and planets appear to move around the sky, not straying far from the ecliptic. If we could travel by spaceship far above the North Pole of the earth, we would look down on the solar system and notice that most objects in it move in a counterclockwise direction. The planets all orbit the sun in a counterclockwise direction and circles are a good approximation to their orbits; most of the planets rotate counterclockwise; and, most of the moons of the planets revolve around their central planet in a counterclockwise direction.

Comets orbit around the sun with periods that range from 3.3 years (Encke's comet) to many million years, and most move in highly elliptical orbits (Figure 1–4). Comets with periods shorter than approximately 30 years (the so-called short-period comets) move more or less like the planets; that is, the planes of their orbits are inclined by small

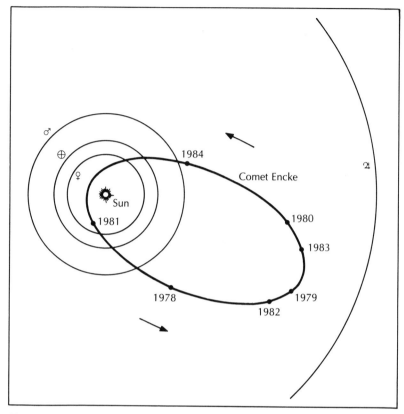

Figure 1–4. Comparison of cometary and planetary orbits as projected on the plane of the earth's orbit. The orbits of the planets are nearly circular and are indicated by their astronomical symbols: ♀ for Venus, ⊕ for Earth, ♂ for Mars, and ♃ for Jupiter. The elongated orbit shown is for Comet Encke. Many comets have orbits much more elongated than the orbit of Comet Encke. Each position is January 1 (drawn to scale).

amounts (not as small, however, as for the planets) to the ecliptic, and they move in a counterclockwise sense as seen from our imaginary spaceship above the North Pole.

By contrast, the motions of comets with periods longer than 200 years (long-period comets) are much more random. The planes of their orbits have all possible angles to the ecliptic, from 0° to 90°, and their motions can be clockwise or counterclockwise. An important contrast to planetary motions, then, is that long-period comets will be observed moving along paths anywhere in the sky. They do not necessarily move along the ecliptic. There is a handful of comets with periods between 30 and 200 years. These intermediate cases must be considered one by one.

CHAPTER 2

HOW TO FIND A COMET

COMET HUNTING

If you find that observing known comets is interesting, you might consider becoming a comet hunter. Your reward for discovering a new comet might be having the comet named for you. Typically, a comet is named for its discoverer. Sometimes more than one observer sights a comet and reports it, and then multiple names can be used. We have already talked about Comet West, which was discovered in 1975 by Richard West of the European Southern Observatory in Chile. Multiple-discovery comets can be given quite cumbersome names. Consider, for instance, Comet Honda–Mrkos–Pajdusakova or Comet Churyumov–Gerasimenko. Simpler naming schemes are also used. For instance, comets are named according to the year of discovery and order of discovery within the year. The first comet discovered in 1973 was Comet 1973a, the second 1973b, and so on. The Czech astronomer Lubŏs Kohoutek discovered two comets in 1973, which were called Comet 1973e and Comet 1973f. In a case such as this, the numbering system is essential to distinguish the two comets, both of which are also called Comet Kohoutek. The scheme does not stop here, however. The discovery order names are merely interim names. Comets are also named according to the year and order in which they pass perihelion; that is, according to the order in which they pass the point in their orbits closest to the sun. Thus the first comet observed to pass perihelion in 1973 was Comet 1973 I, the second to pass perihelion was 1973 II. Comet 1973e became comet 1973 VII and Comet 1973f became Comet 1973 XII. Thus, names with a year and lower-case letter are based on discovery order while names with a year and Roman numeral are based on perihelion passage order. Sometimes a comet may be discovered in one year but pass perihelion in a different year; for instance, Comet 1887c is also Comet 1886 VIII and Comet 1947k is also Comet 1948 I.

How are comets discovered? Through a combination of hard work and luck. This is an area where amateur astronomers contribute as much as professional astronomers. There are a number of amateur astronomers who consistently search for comets. Two things are required for a successful search: a wide-field telescope and a detailed knowledge of the sky. Newly discovered comets frequently are found far enough from the sun that they do not yet have significant tails. They appear as nondescript fuzzy blobs. Unfortunately the universe is full of star clusters, interstellar gas clouds, and galaxies, all of which look like nondescript fuzzy blobs through a small, wide-angle telescope. If you become a comet hunter you must become familiar enough with these blobs to recognize one that seems out of place. Fortunately, the star clusters and gas clouds do not appear to move among the stars, whereas the motion of a comet will be apparent from night to night. If a comet hunter suspects that a fuzzy blob might be a comet, the next step is to watch it for several hours. A moving nondescript fuzzy blob is almost surely a comet. As soon as possible, the observations—including times and the best possible position on the sky—should be communicated to the Central Bureau for Astronomical Telegrams, Smithsonian Astrophysical Observatory, 60 Garden Street, Cambridge, Massachusetts, 02138, U.S.A. Confirmed discoveries are announced on International Astronomical Union postcards (Figure 2–1). The postcards are sent to professional observatories, planetaria, and amateurs in over 50 countries.

Where are comets discovered? Since comets brighten as they move toward the sun, the most productive search strategy is to sweep the sky near the sun in the west after evening twilight and in the east before morning twilight. However, comets can be discovered anywhere in the sky.

An example of a good setup for comet hunting is the observatory built by Leslie C. Peltier, an amateur astronomer from Ohio. He suggests a telescope with a field of view in the 1½° to 2° range. He fitted his telescope with a parfocal turret eyepiece, a mechanism that allows a second eyepiece to be rotated into the field and be in focus. When he sights a suspicious object, he can quickly switch to a higher power eyepiece to examine it. If examination by a higher power eyepiece does not tell what the object is, he resorts to the motion test, using freehand drawings of the object relative to nearby stars. A few hours of observation will usually tell the tale. Peltier's whole observatory rotates on a circular track. He simply sits in a comfortable chair at his eyepiece and sweeps the sky by turning an old automobile steering wheel that operates a belt and pulley mechanism to turn the whole observatory.

Circular No. 3737

Central Bureau for Astronomical Telegrams
INTERNATIONAL ASTRONOMICAL UNION

Postal Address: Central Bureau for Astronomical Telegrams
Smithsonian Astrophysical Observatory, Cambridge, MA 02138, U.S.A.

TWX 710-320-6842 ASTROGRAM CAM   Telephone 617-864-5758

PERIODIC COMET HALLEY (1982i)

D. C. Jewitt, G. E. Danielson, J. E. Gunn, J. A. Westphal,
D. P. Schneider, A. Dressler, M. Schmidt and B. A. Zimmerman report that this comet has been recovered using the Space Telescope
Wide-Field Planetary Camera Investigation Definition Team charge-
coupled device placed at the prime focus of the 5.1-m telescope at
Palomar Observatory. Five exposures of 480-s effective duration
each (in seeing measured to be $1\overset{''}{.}0$ fwhm) were taken on Oct. 16
through a broad-band filter centered at 500 nm. Definite images
near the expected position and having the expected motion of
P/Halley were noted. No coma was detected, and the object had a
Thuan-Gunn magnitude of [g] = 24.3 $\pm$ 0.2 (corresponding to V ~
24.2; and presumably B ~ 25). Two exposures were also made in the
[r] band. Preliminary representative positions, which have an es-
timated external error of $\pm$ $0\overset{s}{.}35$ in α and $\pm$ 5" in δ but greater
internal consistency, follow:

1982 UT	α_{1950}	δ_{1950}
Oct. 16.47569	$7^h11^m01\overset{s}{.}9$	$+ 9°33'03"$
16.49097	7 11 01.8	+ 9 33 02
16.52153	7 11 01.7	+ 9 33 00

The object is located some $0\overset{s}{.}6$ west of the position predicted by
D. K. Yeomans (1981, The Comet Halley Handbook), suggesting that T
= 1986 Feb. 9.3 UT. Confusion with a minor planet would be extremely unlikely. An attempt to confirm the recovery on Oct. 19
was successful in the sense that no objects were detected at the
Oct. 16 locations and that the comet's image would then have been
in the glare of a star; the dense stellar field has in fact
thwarted other attempts to recover the comet during the past
month. The recovery brightness indicates that the 1981 Dec. 18
attempt (cf. IAUC 3688) failed to record the comet by a very small
margin and for an assumed geometric albedo of 0.5 leads to a radius of 1.4 $\pm$ 0.2 km. The comet's heliocentric and geocentric
distances at recovery were 11.04 and 10.93 AU, respectively.

NOVA SAGITTARII 1982

Corrigendum. On IAUC 3736, line 17, the first astrometric
position should be attributed to J. Hers, Sedgefield.

1982 October 21   Brian G. Marsden

Figure 2–1. The International Astronomical Union postcard announcing the recovery of Halley's comet on October 16, 1982. (Courtesy of Smithsonian Astrophysical Observatory, Cambridge, Massachusetts.)

Figure 2–2. Comet Kohoutek (1973f) as shown in the discovery photograph taken on March 7, 1973. The marks in the margin indicate the
location of the comet. (Courtesy of L. Kohoutek, Hamburg Observatory,
West Germany.)

Professional astronomers usually tend to stumble on comets by
accident. The discovery of the two Comets Kohoutek in 1973 is a good
example. Dr. Kohoutek was studying asteroids—small bodies that orbit the sun, primarily between Mars and Jupiter—by photographing
the sky near the ecliptic about 180° from the sun. Both his comets
appeared serendipitously on his photographs (Figure 2–2). Dr.
Kohoutek subsequently verified the comets by obtaining additional
photographs on following nights, and then reported his findings to the
Central Bureau for Astronomical Telegrams.

A discussion of the accidental discovery of comets is not complete without mention of one of the biggest coincidences of all time.
C. D. Perrine was a very prolific comet searcher, with a total of 13
discoveries. In 1896, working at the Lick Observatory near San Jose,
California, Perrine discovered a comet and followed it for a while. At
one point during his observations he received a telegram from a colleague at Kiel, in Germany, reporting the position of the comet. However, there was a slight error in transmission and the telegram reported
an erroneous position, only a small distance from the correct one.
Since the error was small, Perrine didn't notice it. He set up the erroneous coordinates on his telescope, and there was a comet—not the one

he expected, but an entirely new discovery! Once the error was sorted out, Perrine was the proud discoverer of two comets within a few days, a rare occurrence in the life of any astronomer. This is another example of serendipity: the making of worthwhile discoveries by accident. The progress of all science owes much to serendipity.

After the discovery of a comet is announced by the Central Bureau (Figure 2–1), more observations are gathered. Then it is possible to make a preliminary calculation of the size, shape, and orientation in space of the comet's orbit. An expert in these calculations is Brian G. Marsden (Figure 2–3) who is Director of the Central Bureau for Astronomical Telegrams. The parameters resulting from these calculations are used to predict the future motion of the comet and to estimate its brightness.

Figure 2–3. Brian G. Marsden.

FAMOUS COMET HUNTERS

One of the earliest comet hunters has a name that has become well known to amateur and professional astronomers, but not for the dozen comets he discovered. His name is Charles Messier (1730–1817), and he worked in France. As he swept the skies with his telescope searching for comets, he came upon many fuzzy blobs of light that after several night observations proved to be fixed in space. He decided to make a catalog of these objects, so that he and other observers would not mistake them for comets in the future. His catalog, published in 1771, listed over a hundred objects, which we now recognize as the most spectacular examples of star clusters, nebulae, and galaxies in the northern skies. Even today, more than two centuries after Messier, we refer to these celestial marvels by their M numbers—their number in Messier's catalog.

The Herschel family saga makes a fascinating story. Isaac and Anna Herschel lived in Hanover (Germany) in the eighteenth century. Isaac was an oboist in a military band, as well as an intellectual man with an interest in astronomy. His dual interests rubbed off on his children, especially William and Caroline. Caroline (Figure 2–4) is of interest to us because of her success as a comet hunter. However, her story is so closely intertwined with that of her older brother, William (Figure 2–5), that we must look at his life, too.

Around 1760, William Herschel emigrated to England after a disastrous experience as a musician with the regimental band of the Hanoverian Guard. In England he became known as a talented musi-

Figure 2–4. Caroline Herschel, 1750–1848
(Yerkes Observatory photograph).

Figure 2–5. William Herschel, 1738–1822
(Yerkes Observatory Photograph).

cian, music teacher, and composer. In 1766 he became Octagon Chapel Organist at Bath and settled down with considerable financial independence. Several years later, his sister Caroline joined him as his musical assistant. Caroline and William also shared a deep interest in astronomy. Sometime around 1772, William started building telescopes and observing the heavens with Caroline's assistance. Astronomy occupied more and more of their time, and by the end of the 1770s they were spending more time on astronomy than on music.

The turning point in William's career—and as a result in Caroline's as well—came in 1781, when William discovered the planet Uranus. For the rest of their lives, brother and sister devoted their efforts to astronomy. William was eventually appointed astronomer to the King, with Caroline as his assistant. It would be unfair to conclude from this tale that Caroline meekly followed wherever William went. In fact she was a highly intelligent, energetic young woman who shared her brother's enthusiastic interest in the skies.

William Herschel had two positive characteristics that helped him succeed in his astronomical endeavors. He was a superb telescope maker and he was a careful, patient observer. At about the time that he discovered Uranus, he became embroiled in a heated controversy because of the claims he made about the high magnification he used with his telescope. Astronomers of the day were used to instruments that were of limited optical quality, so that high magnification magnified the telescopes' imperfections, not the details of the heavens. The controversy was settled when direct comparisons between William's instruments and those of the Greenwich Observatory demonstrated clearly that William's were superior.

William used his magnificent telescopes to sweep the skies. He pointed the telescope in a fixed direction, and let the rotation of the earth sweep a band of the sky past his view. Sometimes he counted the number of stars in his field of view, other times he recorded the positions of the nebulae (fuzzy blobs) he saw. With his superior telescopes, William increased Messier's hundred nebulae to many thousands.

While William swept the sky for nebulae, Caroline swept for comets. On August 1, 1786 she made her first discovery: the comet known today as Comet Herschel (1786 II). She subsequently discovered, or codiscovered, seven more comets, in 1786, 1788, 1790 (two), 1792, 1795, and 1797. Caroline gained much-deserved recognition in the astronomical world for her work. Not only did she discover comets, she carried out the calculations needed to determine the celestial coordinates of the nebulae that William observed. She continued this work even after William's death in 1822, and in 1823 she was awarded the gold medal of the Royal Astronomical Society, one of the most prestigious honors in astronomy.

Caroline Herschel discovered eight comets in eleven years, with a two-year gap between the first and second discoveries. This shows the enormous patience that a comet hunter must have. It you decide to hunt for comets, you cannot let a year or more of unsuccessful searching deter you.

Jean Louis Pons (1761–1831) may be the all-time record holder. He was the discoverer or codiscoverer of roughly 36 comets between 1801 and 1827. Pons started his career as the concierge at the Marseilles Observatory in France. However, his great success as a comet hunter gave him a boost, and he eventually became director of the Marlia Observatory, near Lucca in Italy.

Clearly, Pons was a man of great patience. Yet, he seems to have been somewhat sloppy in his position measurements. For instance, in February, 1818, he discovered a comet in the constellation Cetus. He was the only observer to see the comet, and his position measurements were too imprecise to permit astronomers to calculate the comet's path through space. In 1873 two observers—Winnecke in Germany and Coggia at Marseilles—discovered a comet almost simultaneously. It was seen for less than a week, but in this case the observations were precise enough to permit the calculation of the shape and orientation of its path in space. Only the comet's orbital period was uncertain. A week of observations is not sufficient to lead to a definitive estimate of any comet's period of revolution around the sun. Nevertheless, the comet's path led observers to speculate that it might be the comet Pons had discovered in 1817. In 1928, A.F.I. Forbes of the Cape of Good Hope (South Africa) discovered a comet that was observed for a sufficiently long time that astronomers could infer not only the characteristics of its orbit in space, but could show that its orbital period was 28 years. It is now clear that Forbes saw the same comet that Pons saw in 1818 and that Coggia and Winnecke saw in 1873. The comet is now called Comet Pons–Coggia–Winnecke–Forbes in honor of all the discoverers.

Astronomy is not really a dangerous occupation. However, Ernst Friedrich Wilhelm Klinkerfues met his death in the line of duty. He was searching for comets at the Göttingen Observatory in Germany on the night of January 28, 1884, as he had on so many other nights. On this fateful night, he fell from the observing platform, broke his neck, and died instantly. During his life, Klinkerfues was a successful comet hunter, with six discoveries between 1853 and 1863, including two in 1854. His first comet (1853 III) was quite spectacular: it became so bright that it could be observed in broad daylight.

Maria Mitchell is considered to be the first woman astronomer in the United States. Nantucket Island, off the coast of Massachusetts, was a Quaker whaling port when Mitchell was born there in 1818. Her father, William Mitchell, maintained a small observatory where he carried out observations required to correct the chronometers that whaling captains used for navigation on long ocean voyages. Mitchell developed a deep interest in astronomy, because, as she said, of a "love of mathematics seconded by [her] sympathy with [her] father's love for astronomical observation."

Mitchell came to the attention of the astronomical world, especially at Harvard University, when she discovered Comet Mitchell (1847 VI). In 1848, she became the first woman elected to membership in the American Academy of Arts and Sciences. She was a member of the faculty of Vassar College when it opened its doors in the fall of 1865, and remained there until she retired on Christmas Day, 1888. She died in 1889. The Maria Mitchell Observatory (Figure 2–6) is still active on Nantucket Island.

Figure 2–6. The Maria Mitchell Observatory and Birthplace, Nantucket, Massachusetts. (Courtesy of the Maria Mitchell Association, Nantucket, Massachusetts.)

Comet hunting can be profitable as well as fun, as the story of Edward Emerson Barnard (Figure 2–7) shows. Barnard was born in Nashville, Tennessee in 1857. He grew up under somewhat difficult circumstances. His father died before he was born, leaving the family with limited means. However, his mother, Elizabeth, raised her son alone and provided him with tutoring and encouragement. As soon as he was old enough, he took a job working for a photographer, a job he held for seventeen years. This work, together with a growing interest in astronomy, set the tone for Barnard's life.

During Barnard's younger years in Nashville, he discovered numerous comets, beginning with 1881 VI, followed by 1882 III, 1884 II, 1885 II, 1886 II, 1886 VIII, 1886 IX, 1887 III, 1887 IV, 1888 V, 1889 I, 1889 III. Here is where the "profitable" factor comes in, for a philanthropic contemporary, H.H. Warner, offered a $200 prize for the discovery of a new comet. Barnard claimed the prize a sufficient number of times to enable him to purchase a home in Nashville with the proceeds. The house was known as "Comet House."

Barnard became one of the premier observational astronomers of his time, and set new standards for the use of photography in as-

Figure 2–7. Edward Emerson Barnard, 1857–1923 (Yerkes Observatory photograph).

tronomy. In fact, the writer of his obituary in the distinguished journal *Monthly Notices of the Royal Astronomical Society* (1923) praised him as "one of the greatest observers of all time, and his work may be compared with that of Tycho Brahe, J.D. Cassini and the Herschels. Like them he combined skill and keenness in observing with a true scientific mind for discovering the things that were important." And, like William Herschel, he had done it all with no formal training.

The tradition of active comet observing has persisted well into the twentieth century. The premier examples among professional astronomers are Georges Van Biesbroeck (Figure 2–8) (1880–1974) who worked at Yerkes Observatory in Williams Bay, Wisconsin, and Elizabeth Roemer (Figure 2–9) who is at the University of Arizona. We have already mentioned two others: the amateur astronomer, Leslie C. Peltier, whose observatory we described in the last section, and C.D. Perrine, who found two comets within a few days. Peltier discovered or codiscovered over half a dozen comets between 1930 and about 1955. One of them, Comet Peltier (1936a), was only ninth magnitude, about 15 times fainter than an object that can be seen with the naked eye.

Most of the men and women we have described here had great success as comet hunters, but not all had formal astronomical training.

Figure 2–8. Georges A. Van Biesbroeck, at the objective end of the Yerkes Observatory 40-inch refracting telescope at Williams Bay, Wisconsin (Yerkes Observatory photograph).

The key characteristics of these observers that enabled them to reach the highest levels of their profession were hard work, care, and infinite patience. There is a group of contemporary comet hunters who are successful practitioners of their trade for similar reasons. They are the Japanese amateur astronomers who spend their time searching for comets.

Minoru Honda, Chief of the Kurashiki Nursery School, has been sweeping the skies for comets since 1937, with considerable success: he has twelve discoveries to his name. Recently, Honda has turned his attention to a search for galactic novae. He makes a wide-angle survey of the Milky Way as often as possible. Comparing the photographs from different surveys, he has discovered five "new" stars or novae. These discoveries have not brought him the attention that his comet discoveries did, however.

Figure 2–9. Elizabeth Roemer.

One of the most interesting modern stories is that of Kaoru Ikeya. Growing up in an economically deprived family, he found it necessary to leave school at the age of 14 to take a factory job polishing piano keyboards in order to help support his family. In his spare time, Kaoru hand ground and polished a 20-cm mirror and built a Newtonian telescope. It took him two years to complete the task. By using second-hand parts, Kaoru finished the telescope at a cost equivalent to a little more than $20. He began a regular program of sweeping the morning sky for comets before reporting for work. For more than a year he searched without success, logging 335 hours of searching on 109 nights. On January 2, 1963, when Kaoru was 19 years old, his persistence paid off: he discovered a twelfth magnitude comet in the constellation Hydra. Before Comet Ikeya (1963 I) had disappeared from view in October 1963, it had reached naked eye brightness in the southern hemisphere skies. The discovery brought the young Kaoru Ikeya unaccustomed visibility to an interested world. He was presented a gold medal by the Astronomical Society of Japan, and his coworkers at the Kawai Gakki Piano Company gave him $300 to help him continue his work.

After the excitement of the discovery diminished, Kaoru settled back into his routine, searching for comets by night and polishing piano keyboards by day. He subsequently discovered Comet 1964f (Comet Ikeya, 1964 VIII) and the spectacular Comet Ikeya–Seki (1965 VIII; see Plate 7), using a new telescope of his own construction. Ikeya's skill at optical work was finally recognized when he was hired by the Norizuki Optical Company to build instruments for professionals. He now manages his own optical workshop.

Tsutomi Seki, who has found several comets, including Comet Ikeya–Seki with Kaoru Seki, came from an affluent family. He is interesting because, like William Herschel two centuries earlier, he makes his living as a professional musician: he plays and teaches guitar.

This brief list of comet hunters—and there have been many others—demonstrates the central role played by amateurs in the discovery of comets. You, too, ought to give it a try!

CHAPTER 3

COMETS IN ANTIQUITY

EARLIEST IDEAS

Once again it is a cool morning before sunrise. A spectacular comet shines brightly on the eastern horizon, which is beginning to be tinted red by the approaching dawn. The tail of the comet sweeps upward away from its head, which lies near the horizon. Imagine that it is eighteen centuries before the Christian era, and we are in Western Asia, at the site of the mighty Babylonian civilization.

We have at our disposal the knowledge our ancestors gained from observing the heavens over many centuries, a knowledge that is quite sophisticated. We have subdivided the sky into constellations,* named after animals, mythical people, and creatures. We know that the celestial bodies rise and set each day, and that the stars rise a few minutes earlier each night, causing a gradual shifting through the year of the constellations that are visible at any given time. Cygnus the Swan is visible in the summer sky, Taurus the Bull in winter.

Our studies of the sky started out merely to mark the passage of the seasons, to predict the time for the planting of crops, and to tell the times of holidays. As time progressed, however, the observations have been used more and more for astrological reasons: to predict human events from the positions of the stars and planets.

We Babylonians have charted the slow motion of the known planets—Mercury, Venus, Mars, Jupiter, Saturn—as well as the sun and the moon relative to the fixed stars. We know that Mercury and Venus never stray far from the sun. One or the other can be observed rising before the sun in the morning sky or setting after the sun in the evening sky. We know that Venus is at most one-eighth of a full circle away from the sun (more precisely, 47°), and moves from the morning sky, past the sun to the evening sky, and back to the morning sky in

*Some, but not all, of the Babylonian constellations were the same as our modern constellations.

584 days. These planets always lie near a path around the sky we call the ecliptic.

We know that the motions of Mars, Jupiter, and Saturn are similar to one another, but are considerably different from those of Mercury and Venus. First, the former three planets can be situated anywhere on the ecliptic, and are not necessarily found near the sun. Normally, these three planets move slowly eastward among the stars, never straying far from the ecliptic. Second, periodically each planet will stop its eastward motion, make a small loop in the sky (which takes a month or so), and then resume its eastward motion. Mars makes this retrograde loop roughly every two years; Jupiter and Saturn every 13 months or so. Actually, we know all these times very precisely. We have studied these planetary motions in detail, but we have only vague ideas of their cause. We don't know what planets are, other than wandering lights in the sky.

Now we are observing a comet in the Babylonian skies. It does not behave like the planets—it does not lie near the ecliptic and it has an extensive tail. What can it be? What can it mean? Last year, when the crops failed, a bright comet appeared in the sky. Perhaps this comet is a portent of yet another disaster. Our wonder and awe at the beauty of the comet is tinged with superstition and fear.

Looking back at the science of Mesopotamia from our twentieth-century vantage point, we see a lot that was impressive, particularly in the achievements of the late Chaldean culture. Their tables of the motions of the planet Venus (their *Ishtar*) are truly remarkable. Yet, in known Mesopotamian texts there are almost no references to comets. Much of the writing must have been lost, for Greek and Latin thinkers do report on Babylonian comet lore.

Comets were also observed in prehistoric times in the western hemisphere (Plate 2), and were mentioned in sixteenth-century reports of Aztec Indian records (Figure 3–1).

The history of the development of concepts about comets is an example of the development and application of the scientific method. The first step in science is to observe things, to see how a certain part of nature behaves. The second step is to experiment, and the third is to form a theory. Our ancestors carried out the first step very well.

Over the centuries since the time of Babylon, scientists have attempted to answer the question, What is a comet? To trace the development of our modern knowledge, we must weave together many attempts to understand the nature of the planets and of comets. The earliest debate, which raged throughout scientific discussions for

Figure 3–1. Aztec representation of a comet, from a native description of the conquest of Mexico in 1519 and 1520.

over a thousand years, was over the question, Are comets celestial bodies or atmospheric phenomena?

We will begin our story at the time of the great Greek philosophers. The Greeks added the missing element to Babylonian thought in that they attempted to give geometrical explanations of celestial motions in terms that added a predictive element. That is, with their models the Greeks could predict the future positions of the planets.

Plato (428–347 B.C.) founded a school, "The Academy," in 387 B.C. on land near Athens that once belonged to the Greek hero Academos. (By this one act, incidentally, Plato guaranteed that Academos' name would become a familiar word in virtually every language with European roots.) Plato himself did very little writing on astronomy, but he stimulated his students to do so. One of the first of these was Eudoxus of Cnidos. Eudoxus imagined a set of 33 spheres, all centered on the earth, interconnected in such a way that the rotation axis of any sphere was attached to the surface of the next larger sphere at some predetermined angle to the axis of that larger sphere. Each sphere in the complex rotated with a certain speed. By choosing the orientations and speeds of the spheres appropriately, and by attaching each planet to the appropriate sphere, Eudoxus came up with a model that mimicked the motions of the planets very well. Eudoxus himself did not propose that the spheres represented reality. For him they were merely a convenient mechanism—like our modern mathematical description of planetary motions—for predicting the future positions of the planets.

Perhaps the most famous student at the Academy was Aristotle (384–322 B.C.), who came to study with Plato in 367 B.C. and remained for twenty years. We know a great deal about his life, including the fact that he was hired by Philip of Macedonia to tutor Philip's young son Alexander. Aristotle's writings are encyclopedic, including works on mathematics, logic, astronomy, physics, geography, geology, biology, and medicine. Reading Aristotle's works on astronomy, we can see his disagreement with Eudoxus' concept of homocentric spheres. Not only did Aristotle add an additional 22 spheres, bringing the total to 55, but he strongly implied that the spheres were real objects, not a convenient fiction.

Comets are discussed in Aristotle's book on meteorology, not astronomy. He regarded both comets and the Milky Way as phenomena in our atmosphere. He first reviews the thoughts of some of his predecessors. The followers of Pythagoras—the discoverer of the famous theorem about right triangles—believed that a comet was a planet that was seen above the horizon only on rare occasions. Others of the same era believed comets to be celestial bodies. Aristotle disposed of these ideas by pointing out that the other celestial bodies, the planets, all move in a predictable fashion along the ecliptic while comets appear anywhere in the sky at erratic times. They behave in a very unplanet-like way.

Aristotle begins the discussion of his view of the nature of comets with a statement of his philosophical viewpoint: "We consider a satisfactory explanation of phenomena inaccessible to observation to have been given when our account of them is free from impossibilities."

Aristotle theorized that all matter was composed of four "elements": fire, air, water, and earth. These elements can originate from one another, and each has the others contained within it (Figure 3–2). When the sun warms the earth, the earth evaporates into two types of vapors, a moist, cool one and a warm, dry, windy exhalation. The windy exhalation then rises above the moist vapor and reaches the edge of space or, as Aristotle put it, "the outermost part of the terrestrial world which falls below the circular motion." The circular revolution carries the warm exhalation around the earth. "In the course of this motion it often ignites wherever it may happen to be of the right consistency, and this we maintain to be the cause of . . . [shooting stars]." A comet is formed when the circular motion introduces into the warm exhalation just the proper level of fire—not too little to be merely a shooting star, and not too much to ignite everything (Figure 3–2). "The kind of comet varies according to the shape which the exhalation happens to take. If it is diffused equally on every side the star is said to be fringed; if it stretches out in one direction it is called bearded."

Aristotle went on to argue in favor of his model by stating that the

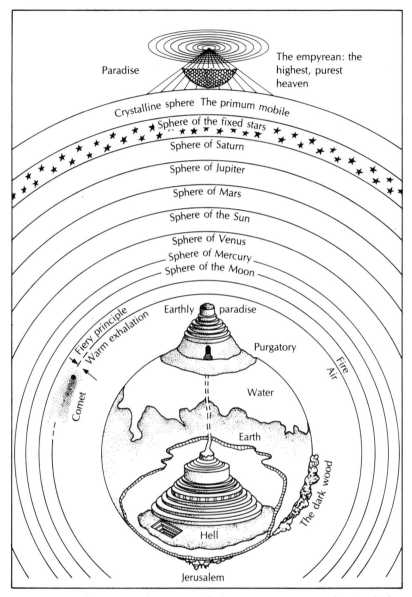

Figure 3–2. The Aristotelian view of the universe as "modernized" by Dante in The Divine Comedy. *Dante's Hell, shown in cutaway, is inside the earth. We have simplified the figure and added the Aristotelian theory of comet origin: Comets were supposed to be formed when the fire interacts with a warm exhalation from the earth.*

presence of a comet is an indication of coming wind and drought. He argued that this fact clearly shows the "fiery constitution" of comets and shows they arise from drier air. He describes the Milky Way in a similar way.

We must not be too quick to scoff at this theory. It is quite consistent with the picture of the world held by philosophers at that time. And, as Aristotle said, that was the real test of a hypothesis.

After the time of Aristotle, a number of Roman authors discussed comets in extensive treatises. One of the most thorough students of comets was Seneca (4 B.C.–64 A.D.), who wrote *Questiones Naturales*. In reply to Aristotle's argument that comets cannot be planetary bodies because they move outside the ecliptic or zodiac, Seneca states, "If [a comet] were a [planet], says some one, it would be in the zodiac. Who, say I, ever thinks of placing a single bound to the stars? Or of cooping up the divine into narrow space? These very stars, which you suppose to be the only ones that move, have as everyone knows, orbits different one from another. Why, then, should there not be some stars that have a separate distinctive orbit far removed from them?" Seneca is quite correct, but his argument is based on logic, not scientific measurements.

Geometrical methods started by Plato's students for predicting the future positions of the planets were brought to a high level of accuracy by Claudius Ptolemaeus, or Ptolemy (ca.100–ca.178 A.D.). The details of his so-called epicycle theory are not important here. What is important is the fact that this highly flexible theory provided predictions of the motions of the planets that were as good as the observations that could be made at the time. As a result, there was little need to attempt to improve the model. For the next twelve centuries, the theory of planetary motions was considered to be a solved problem.

MEDIEVAL COMET LORE

From the fall of the Roman Empire—which occurred soon after the attacks of Attila the Hun and his hordes around 450 A.D.—to the Renaissance, science was kept alive in the Middle East. Very little that was new or interesting was written in the West about the nature of comets, or anything else, in this period, partly as a result of the success of Ptolemy's epicycle theory. However, records of bright comets are scattered here and there in medieval European writing, and an occasional work about the nature of comets does exist.

One example comes from a period in the seventh and eighth centuries known as the "Northumbrian Renaissance." A number of monasteries were founded by Benedictine and Irish missionaries in the area of England now known as Northumberland. Two of these monasteries were founded by Benedict Biscop (now known as St. Benedict Biscop), who had lived in Rome as a young man. When he built the monasteries of Peter and Paul at Wearmouth and Jarrow, he provided them with excellent libraries of manuscripts he had brought from the continent. Included in the libraries were the classical works on comets by Seneca and by Pliny.

In about 680, a seven-year-old boy was entrusted to Abbot Benedict for his education. The young boy, Bede, grew up to be one of the most notable intellects of the time. He remained in the monasteries his entire life, studying the manuscripts in the library and writing on many subjects, including comets. He is known to have written almost three dozen books, two of which, *De Rerum Natura* (the Nature of Things) and *Historia Ecclesiastica* (A History of the English Church and People) touch on comets. In both of these books it is clear that Bede saw comets as portents of unusual happenings such as the death of kings, war, pestilence, and other dread occurrences.

In the *Historia Ecclesiastica,* which modern historians consider to be one of the best existing sources for English history up to 735, Bede talks about several comets. Two bright comets appeared in January, 729, and remained visible for about two weeks, one in the evening sky and one in the morning sky. Bede states that the two comets on opposite sides of the sky "portended awful calamity to east and west alike. Or else, since one comet was the precursor of day and the other of night, they indicated that mankind was menaced by evils at both times." Bede then describes the results of the cometary appearance: "The Moors ravaged Gaul with a horrible slaughter," then were defeated by Charles Martel at the Battle of Tours (732); and King Osric of Northumbria died leaving his throne to Ceolwulf. "Both the onset and course of Ceolwulf's reign were filled by so many grave disturbances" that Bede would not predict in 735 what the outcome would be. (King Ceolwulf laid down his crown in 737 and became a monk.)

The great monasteries of Northumbria were destroyed by the Vikings at the end of the eighth century. St. Bede the Venerable was canonized in the nineteenth century.

There is a significant list of other writers of the Middle Ages and early Renaissance who dealt with comets, including Thomas Aquinas, Roger Bacon, Regiomontanus, and others, but they are primarily of interest to the historian of science. Some of the most exciting and beautiful records of comets appear in the art of the era. In Chapter 9 we will describe the events surrounding the appearance of Halley's comet in 1066. That appearance was celebrated in the well-known Bayeux Tapestry (Plate 3).

The life of Giotto di Bondone (ca.1276–ca.1337) spanned the time when Italy was moving into the full flower of the Renaissance. Born to a peasant family, Giotto was a precocious artist who was discovered at an early age by the Florentine artist Cimabue. Under the tutelage of Cimabue, Giotto developed into one of the great masters of his day. Among his greatest masterpieces are the frescos he painted on the walls of the Scrovegni Chapel in Padua.

Reginaldo Scrovegni, history says, was an evil man—so evil that Dante specifically mentions him as among the usurers in the Inferno. To atone for his father's wickedness, Reginaldo's son Enrico built a chapel in Padua and commissioned Giotto to decorate its interior walls with frescos. The little chapel, completed in 1303, is a magnificent example of artistic genius. One of the scenes, the Adoration of the Magi, shows the three "wise men" presenting their gifts to the Christ Child. What is interesting in the fresco is the star of Bethlehem: it looks very much like a comet (Plate 4). Halley's comet would have been visible in European skies in late 1301, and was probably a spectacular sight. Art historians speculate that Giotto was so impressed by the comet that he painted its portrait as the Christmas star. If that is true, this is certainly one of the most magnificent of all the records of the comet.

Giotto's fresco is not the only comet portrait known to us. Albrecht Dürer (1471–1528) depicts a comet in his engraving *Melencolia I*. If we accept one possible definition of Medieval times as the era from Constantine to Columbus (ca.300–ca.1500), then Dürer's engraving, dated 1514, is an early Renaissance portrait of a comet.

THE RENAISSANCE

The stage for our modern understanding of the planets was set in the sixteenth century with the publication of Nicolas Copernicus' (Figure 3–3; 1473–1543) book, *On the Revolutions of the Heavenly Spheres*. In this book he states that contrary to the Aristotelian model, the sun is the center of the universe, and all the planets including the earth move around it. The daily rising and setting of the stars is due to the earth's rotation, and the motion of the sun among the stars is merely a reflection of the earth's circular motion around the sun (Figure 3–4). Copernicus caused a flap among the scholars of the day, who accepted the word of Aristotle as a matter of faith. Copernicus' model of the solar

Figure 3–3. Nicolas Copernicus, 1473–1543 (Yerkes Observatory photograph).

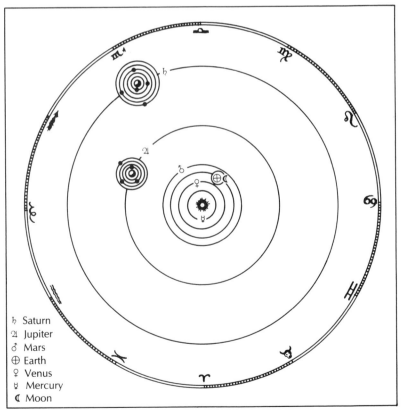

ħ Saturn
♃ Jupiter
♂ Mars
⊕ Earth
♀ Venus
☿ Mercury
☾ Moon

Figure 3–4. The Copernican universe with the sun at the center.

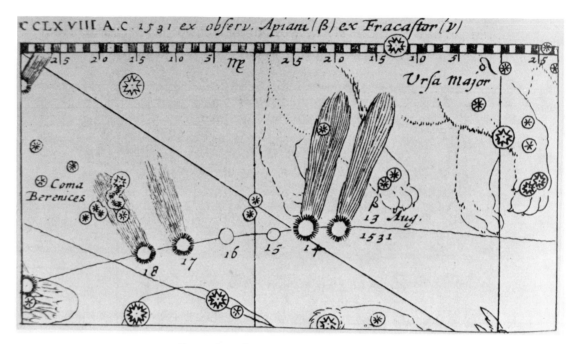

Figure 3–5. Peter Apian's observations of Halley's comet in 1531. The observed orientation of the tail showed that the tails pointed away from the sun (Bulletin de la Société Astronomique de France, *1910*).

system had little scientific basis, other than some aesthetics, when it was first proposed; he objected to the complex series of epicycles developed by Ptolemy. However, that would change in the century after the publication of Copernicus' theory, largely due to the work of three men: Tycho, Kepler, and Galileo. These three brought the observational stage of the scientific method to its culmination, and Galileo ushered in the experimental stage.

Some cometary research was being carried out at the time of Copernicus. For instance, the Italian astronomer Fracastoro (ca.1480–1553) reported observations of comets in his *Homocentrica*, and stated that their tails always point away from the sun. Contemporaries of Fracastoro, such as the German astronomer and mathematician Peter Apian, also noted the same fact (Figure 3–5).

Copernicus wrote his great treatise in the waning days of the Renaissance. The debate on the theory began slowly but gathered momentum over the years. It broke out into open warfare in the "Age of Reason" (1550–1650). The Elizabethan age was in full flower in England. William Shakespeare was in his prime. Thomas Digges (ca.1545–1595) was an ardent defender of Copernicus' theory. He pictured a universe infinite in extent and populated with stars, some of which were so far away as to be invisible. Francis Bacon (1561–1626) argued against the theory. In France, René Descartes (1586–1650) defended Copernicus from every lecture platform available to him. In Italy, the debate divided the church fathers. Many, like Giordano Bruno (1548–1600), argued forcefully in favor of Copernicus. Bruno was way ahead of his time. Like Digges, he believed in an infinite universe. Most of the debate was on philosophical grounds. Aesthetics, beauty, and logic were the key words. But that was changing. The acceleration of development toward modern ideas began in the late years of the sixteenth century, over 400 years ago.

CHAPTER 4

MODERN IDEAS DEVELOP

During the Renaissance, the rebirth of thought about ourselves and the universe we live in set the stage for the development of modern science. Following the publication of Copernicus' *De Revolutionibus* in 1543, many important scientists contributed to changing our view of the heavens. As the world view changed so did the view of comets. Our modern concepts of comets (and the universe) are based on modern, high-quality observations. It is natural then, to begin the story with one of the great observers of all time, Tycho Brahe.

TYCHO BRAHE

In 1546, Tycho Brahe (Figure 4–1) was born to a noble Danish family. Early in his life he was abducted by a childless rich uncle, who then convinced the parents to let him keep the boy. The event proved advantageous to Tycho. He was educated at the best universities in Europe, the University of Copenhagen and the University of Leipzig. When the uncle died in 1565, he left his sizable fortune to Tycho.

There is an amusing and revealing anecdote about Tycho. The impression we have of him is that he was an arrogant, quick-tempered young man. Soon after his uncle died, Tycho got involved in a dispute that led to a duel, in which his nose was partially cut off by his opponent. History says that Tycho had a gold replica made of his nose, which he wore the rest of his life. Recently, Tycho's grave was moved to make way for progress, a new superhighway in Prague. His coffin was opened, partly out of curiosity. Would the gold nose be there? No gold nose was found, only a tell-tale green stain over the nasal opening of the skull. It appears the fake nose was, after all, a baser metal.

In 1572, a new star appeared briefly in the constellation Cassiopeia—a phenomenon we know today as a supernova—throwing

Figure 4–1. Tycho Brahe, 1546–1601 (Yerkes Observatory photograph).

scientists of the day into a tizzy. Aristotle had argued that the heavens were unchanging. Few, however, could argue with the evidence of their own eyes. Tycho was one of the first to notice the star. With the widespread recognition the discovery brought him, he approached Frederick II of Denmark in an attempt to interest him in building an observatory. Frederick gave Tycho the island of Hveen as a fiefdom, and granted him an income. There Tycho built the observatory Uraniburg, and outfitted it with the best instruments obtainable. A quarter century before the telescope was invented, Tycho set about measuring the changing positions of the planets with the highest precision possible. Tycho's instruments consisted of quadrants, sextants, and other devices for measuring the positions of stars and planets on the sky. Despite the fact that the instruments had no optical components, the best of them could be read to 10'' (the diameter of a dime seen at 300 meters). At last the data on which scientific arguments could be based were being obtained. Tycho continued his measurements until 1597. By that time Frederick II had died and the new regime had little sympathy for the astronomer's extravagances. However, Tycho sold his services to Emperor Rudolf II, as court astrologer. He moved to Prague and advertised for an assistant to help reduce his data. A young man named Johannes Kepler (1571–1630) came to Prague and spent years working on the data, first with Tycho, then alone.

Tycho died in 1601. He was at a dinner attended by Rudolf II. No one could leave the table until the Emperor left—that would have been a breach of etiquette of the worst sort. Not being able to heed a call of nature, Tycho suffered a burst bladder and died from complications.

During his life, Tycho made one major contribution to the study of comets: as a result of his observations of the comet of 1577, he proved that comets are celestial objects.

THE COMET OF 1577

In November, 1577, a bright comet appeared in the sky, and it continued to be visible until early 1578. Tycho, at that time still in Copenhagen, and his contemporaries throughout Europe made measurements of the position of the comet with respect to the stars. One of the most important of these contemporaries was Hagecius (ca.1526–1600) who observed from Prague.

Both men attempted to measure what we call the comet's *geocentric parallax* (Figure 4–2). The comet's position is measured when it is high and low in the sky, and is thereby seen from two vantage points whose separation is roughly the radius of the earth. If the comet is nearby, it should shift its position relative to the stars. A measurement of the angular shift permits a calculation of the comet's distance from the earth. Tycho had already measured the parallax of the moon in this way and found it to be roughly 1°. His observations of the comet revealed a parallax only one-forth as large (approximately 15′), placing the comet at least four times the moon's distance from us. At the same time, Hagecius announced a parallax of 6°, which would place the comet very near the earth. Tycho was eventually able to show that Hagecius had misinterpreted his measurements. Hagecius' observation in fact agreed with Tycho's. In addition, Tycho could compare his and Hagecius' observations and obtain a parallax with a Copenhagen-Prague baseline.

Once Tycho knew the comet's distance, he could calculate its dimensions. He measured the tail to be 22° long and 2½° wide and the head to be 8′. A parallax of 15′ corresponds to a distance of roughly 1 million km. Thus Tycho's measurements meant the tail was over 350,000 km long, and the head was over 2000 km across, clearly very large numbers. The comet of 1577 was a huge celestial object, not an atmospheric exhalation. Tycho felt that this conclusion was probably true for all comets.

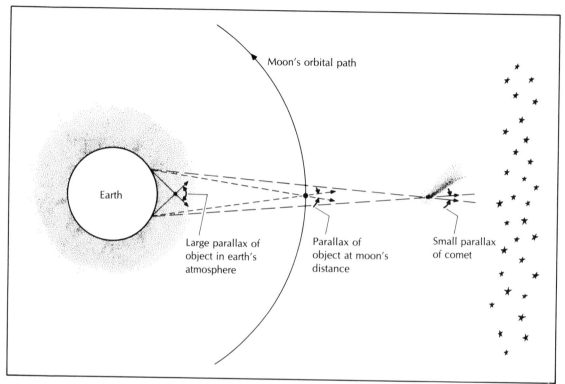

Figure 4–2. Diagram illustrating the concept of geocentric parallax (not drawn to scale). The parallax angle is marked for three different distances from the earth. Notice that this angle is smaller for distances farther from the earth; that is, the separation in angle on the sky is smaller for objects farther from the earth. By measuring this angle for the comet of 1577, Tycho Brahe established that the comet was well beyond the moon.

JOHANNES KEPLER

Johannes Kepler (Figure 4–3) was born near Stuttgart in 1571. His father, an army officer, was not a man of means. However, the Duke of Württemburg paid for the young man's education. At the University of Tübingen, Kepler became a convinced Copernican. After completing the university course, Kepler settled in Graz, Austria, as a schoolmaster, where he married and had a houseful of children. He was a Protestant, however, and was forced from Graz by Archduke Ferdinand. Just when things seemed bleakest, Kepler learned of Tycho's need for an assistant in Prague and took the job.

Kepler began work in Prague at one of the low points in the

Figure 4–3. Johannes Kepler, 1571–1630 (Yerkes Observatory photograph).

history of science. In 1600, Giordano Bruno was burned at the stake in Rome as a heretic, and as a result many supporters of the Copernican picture of the universe decided discretion to be the better part of valor and kept quiet. René Descartes, for instance, never again mentioned the theory.

When Tycho died, Kepler was appointed his successor and given a salary by the Emperor. However, Kepler was not given access to Tycho's instruments, only to his records. His only recourse was to make a detailed study of the existing data. Over the following years, Kepler formulated three fundamental laws of planetary motion, based on Tycho's data. He began by showing that the orbit of Mars around the sun is an ellipse with the sun at one focus. By the process of inductive reasoning, arguing from the particular to the general, Kepler concluded that all the planets move in elliptical orbits. Tests of this hypothesis all proved positive. Kepler also noted that planets move fastest in their orbits when closest to the sun and slowest when farthest away. His detailed studies produced the first two so-called "laws" of planetary motion:

1. All planets move in elliptical orbits with the sun at one focus.
2. The straight line joining a planet to the sun sweeps out equal areas in equal intervals of time.

These two laws apply equally well to comets. However, while the orbits of the planets are nearly circular, cometary orbits are very extreme ellipses, approaching parabolas in many cases. It is interesting to note, however, that Kepler did not recognize this fact. He argued that comets move in straight lines at irregular speeds. Here, Tycho was ahead of his younger colleague, because he postulated a circular orbit near the orbit of Venus for the comet of 1577.

GALILEO

One of the superstars in the story of our growing understanding of the solar system, and therefore of comets, was Galileo Galilei (Figure 4–4). He was born in Pisa in February, 1564. In adult life, he became a faithful supporter of the theories of Copernicus, undaunted by the fate of Giordano Bruno. Galileo's most famous accomplishment was his use of the telescope to study the heavens. With it, he made a fantastic array of discoveries, one of which was that Venus, like the moon, exhibits phases. If Venus orbited the earth as the Greeks predicted, then it would not show phases. However, if it orbited the sun as Copernicus asserted, then it would exhibit phases. Thus, Galileo's single discovery of the phases of Venus was compelling proof that Copernicus' view of the solar system was correct.

Figure 4–4. Galileo Galilei, 1564–1642 (Yerkes Observatory photograph).

The story of how Galileo ran afoul of the authority of the Church is a fascinating chapter in the history of science, which we will not recount in detail here. Galileo, the true scientist, offered his telescope to the disbelievers. "Here," he would say, "look for yourself, the evidence is incontrovertible." His detractors would reply, "We need not look, we have faith that you are wrong; the evidence you claim can't be there." It is amazing to think that there were people so convinced of their ideas that they would not confront the facts—or were they really very insecure in their positions? In the short run Galileo was silenced by the Inquisition; in the long run the truth of his viewpoint prevailed.

Three comets appeared in 1618, including a very bright one. They sparked a heated controversy between Galileo and several of his contemporaries. Galileo argued that comets might be optical illusions, in which case Tycho's parallax measurements were meaningless. Galileo returned to the old exhalation theory. This controversy led him to publish his lengthy letter "to the Illustrious and Very Reverend Don Virginio Cesarini," *The Assayer.* The cometary physics in that work is flawed. However, the scientific philosophy Galileo exposes is highly modern in tone. The work is amusing to read, for Galileo deals with his detractors in what can only be described as nasty digs.

When we look at Galileo's over-all life work, we can forgive him for his backward view of comets. Galileo is also usually credited with the introduction of experimentation, the second step in the developing scientific method. Until his day, scientific progress was made as individuals collected increasingly precise observations (supplemented by a few experiments, by men such as Aristotle). Galileo's experiments were largely in the field of classical mechanics. For instance, his experiments with falling bodies (which legend says he dropped from the Leaning Tower of Pisa, but which actually used inclined planes, in a slow-motion version) are justifiably famous. Theses experiments laid the groundwork for the theory of classical mechanics that Newton developed to its modern level, ushering in the third stage of science: the theoretical stage.

NEWTON AND HALLEY

Isaac Newton (Figure 4–5) was born on Christmas Day, 1642, the same year Galileo died. Newton was one of the great intellects of all time. It is a strange coincidence that Galileo was born the day

Figure 4–5. Isaac Newton, 1642–1727
(Yerkes Observatory photograph).

Figure 4–6. Edmond Halley, 1656–1742.

Michelangelo died and Newton the year Galileo died. It is almost as if the world couldn't cope with too many minds of this level, and had to release one before the next came along.

Newton's contributions to science are many, in mathematics, optics, astronomy, and religion. Interestingly, he thought religion would be the subject in which he would make his mark, but his thoughts on the subject are all but lost today. From our point of view, the most notable of his contributions are the laws of motion and gravitation. Based on his studies, Newton developed a quantitative law of gravitation that said, among other things, that every body in the solar system attracts every other body. The precise mathematical form of the attraction can be inferred from Kepler's laws of planetary motion. The massive sun holds onto its retinue of planets, comets, and other lesser bodies because of gravity. In addition, the planets exert small tugs (or perturbations) on one another that cause measurable departures from exact elliptical motion. Jupiter, the second most massive object in the solar system, is responsible for the majority of the perturbations.

Edmond Halley (1656–1742; Figure 4–6) was fourteen years younger than Newton. Like Newton, he was a man of superior intellect. Newton was always reluctant about publishing his ideas, and Halley saw that many of them got into print. Newton's great *Mathe-*

matical *Principles of Philosophy* or *Principia* would never have seen the light of day had it not been for Halley's efforts to get it published. Halley himself contributed much to the world of science. For instance, he compiled the first actuarial tables ever published. His greatest contribution, however, has to do with the comet that bears his name.

By the late seventeenth century it had become well accepted that many comets move in parabolic, or nearly parabolic, orbits. Newton developed a method to calculate the size and the orientation in space of a parabolic orbit using three observations of a comet's position among the stars. Halley used this technique to calculate the characteristics of the orbits of a number of well-observed comets. One of the best examples was the great comet of 1680. Halley demonstrated that that comet moved in a very elliptical orbit that differed little from a parabola when the comet was near the sun. He went on to study roughly two dozen other comets, including a bright one that appeared in 1682. He immediately found an interesting result: the comets of 1531, 1607, and 1682 all seemed to move in the same orbit (Figure 4–7). The orbit was similar to that of a more poorly observed comet that had appeared in 1456. Halley concluded that these were all the same comet that orbited the sun with a period of roughly 76 years. He predicted a return in 1758, based only on the interval between previous appearances. However, he was well aware that Jupiter and even Saturn could perturb the motion of the comet and alter its period somewhat.

The French astronomer A. Clairaut (1713–1765) calculated the influence of the major planets on the comet and predicted that it would pass perihelion—the point closest to the sun in its orbit—on April 15, 1759. The comet was recovered Christmas night, 1758, and passed perihelion on March 13, 1759, a time well within the uncertainty of Clairaut's prediction.

This one comet did more than anything else to dispel any doubts about Newton's mechanics. The modern heliocentric picture of a solar system bound together by gravitation could no longer be doubted. To honor Halley, the comet bears his name, one of only three comets that are named not after discoverers but after mathematicians who studied their motion. Another, Encke's comet, will be discussed in Chapter 8.

An amusing sidelight to the confirmation of Newtonian mechanics is the fact that Halley's comet passes through the distances of Venus, Earth, Mars, Jupiter, Saturn, Uranus, and Neptune in its passage through the solar system. If Aristotle were correct that a set of concentric crystalline spheres controls the motions of the planets, the comet would certainly collide with them!

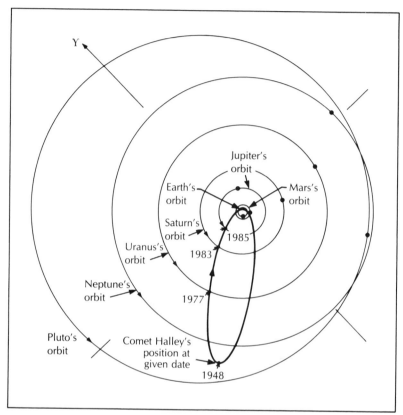

Figure 4–7. The orbit of Halley's comet. (Courtesy of D.K. Yeomans, from The Comet Halley Handbook, *The International Halley Watch, NASA/Jet Propulsion Laboratory, Pasadena, California, 1981.)*

THE SCIENTIFIC METHOD

Today physical science advances by the same three steps: observation of a natural phenomenon; experimentation to see how the event behaves under controlled conditions (astronomy is an exception in physical science—it is difficult to experiment on celestial bodies); and, finally, development of theories, usually based on mathematics, to explain the phenomenon. In the process of developing theories, scientists still follow the rules of reasoning set down by Newton in the *Principia*. The most important of these rules is to keep the theory as simple as possible; in Newton's words, "We are to admit no more causes of natural things than such as are both true and sufficient to explain their appearances." In general, the mathematical theory of a phenomenon is a very succinct description of the results of all observations and all experiments that went before. More important, however, a good theory often leads to the prediction of new relationships among phenomena.

In the preceding discussion we have taken brief glimpses at the scientific method. Scientists become aware of events in nature by observing them. Researchers then proceed to gather quantitative measurements of the phenomena. For instance, Tycho's measurements of the parallax of the comet of 1577 fall into this observational stage. Experimental science, as practiced by Galileo, led to classical mechanics, which has given us an understanding of the motions of comets, but until we send space probes into comets (Chapter 11) we have little hope of carrying out direct experiments on them. Theoretical science was begun by Newton, who developed the mathematical tools (calculus) and applied them in a logical manner in developing his theory of planetary motions.

PRESENTING COMETS TO THE PUBLIC

Let us now turn from technical matters and take a look at how astronomy and comets were presented to the public in the late Renaissance and the early modern era. When movable type found its way to Europe in the mid-fifteenth century, an interesting publishing form became popular: the broadsheet. The broadsheet was basically a poster, a single sheet of paper printed on one side, discussing an event or topic of current interest. The printing technology of the day was limited and press runs seldom exceeded a few thousand copies. The sheets were sold on the streets, much as newspapers are today. Of the broadsheets remaining today, the most interesting to us are the slightly over 200 that deal with astronomical topics, and particularly those that deal with comets. The earliest known astronomical broadsheet was published in 1492, as the Renaissance was opening. It discusses a meteorite that fell in Alsace in November of that year. A number of later broadsheets presented information on Tycho's supernova of 1572 to the public, some claiming the star to be a comet. A fine example of a nineteenth-century broadsheet about comets is shown in Figure 4–8.

Figure 4–8. Broadsheet announcing the Great Comet of 1843 (see also Figure 8–4). (Phot. Bibl. nat. Paris.)

CELESTIAL MECHANICS

The return of Halley's comet in 1759 firmly established Newtonian mechanics as the physical system to use in predicting motions in the solar system. Celestial mechanics was developed in the remainder of the eighteenth and nineteenth centuries into an elegant, formal, mathematical system.

From the point of view of cometary physics, the problem of celestial mechanics can be stated quite succinctly in three steps. The first step, after a comet has been discovered, is to ascertain the path it follows through space, given a series of observations of its position on the sky. The second step is to predict the future path of the comet on the sky, including the influence of the perturbations of the planets. And the third step is to calculate the path the comet must have followed before its discovery.

When a comet is discovered, it is moving in an elliptical orbit, which may be nearly parabolic for comets with very long periods. In the long run, planetary perturbations will cause the orbit to change slightly. However, at any instant the orbit can be considered an ellipse. If the planets were all to disappear at some instant, the comet would continue to orbit the sun in the ellipse characteristic of its motion at that instant. The orbit the comet would follow if planetary perturbations were magically turned off at some time is called the *osculating orbit* for that time.

How is an ellipse described? First, we must specify its size and shape. To do so, we specify the semimajor axis, a, and eccentricity, e, of the ellipse. Instead of the semimajor axis, for nearly parabolic orbits astronomers sometimes quote the perihelion distance, q, which is the distance from the sun to the point on the cometary orbit nearest the sun. The semimajor axis, perihelion distance, and eccentricity are simply interrelated. Once the size and shape of the orbit are given, its orientation in space must be specified. Here, astronomers often use three angles (Figure 4–9): the inclination of the orbital plane to the plane of the ecliptic, i, the longitude of the ascending node, Ω, and the argument of perihelion, ω. The ascending node is the point on the orbit where the comet passes from below the plane of the ecliptic to above the plane of the ecliptic. These three angles uniquely specify the orientation of the orbit in space. (We need not get involved in the details of precisely what these angles are to understand the basic concept.) Five parameters, called *orbital elements, a, e, i, Ω,* and ω (or q, e, i, Ω, and ω) completely specify the size, shape, and orientation of the orbit. Now all we need to know is where the comet is along the orbit. As the final orbital element, astronomers usually specify one date when the comet passes the perihelion point on its orbit. The formalism of celestial mechanics gives us the mathematical tools to calculate where the comet is in its orbit at any time, given one time of perihelion passage.

Given where the comet is in its orbit and the orientation of the orbit in space, we can calculate the comet's position in space at any instant. We also know a great deal about the earth's motion around the sun, and we can calculate its position in space at the same instant. Knowing these two points in space permits us to calculate where the comet will appear to be on the sky as seen from the earth.

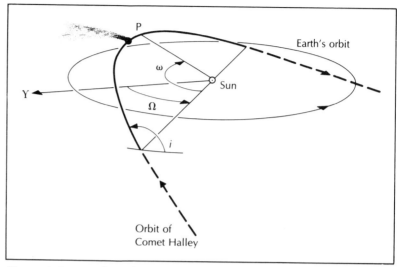

Figure 4–9. Angular orbital elements for the case of Halley's comet. (Courtesy of D.K. Yeomans, from The Comet Halley Handbook, The International Halley Watch, NASA/Jet Propulsion Laboratory, Pasadena, California, 1981.)

The problem that celestial mechanicians attacked in the late eighteenth and early nineteenth century was the inverse of the problem we just described. Given that we have made a series of observations of a comet's position among the stars, as seen from the moving earth, we wish to find the elements of its orbit. One simplification that is sometimes used as a starting point is to assume that the orbit is exactly a parabola. For this case, the eccentricity is $e = 1$ by assumption, and we need find only five additional elements. Interestingly, an effective solution to this problem was published in 1797, not by one of the great celestial mechanicians but by an amateur astronomer who was a physician by trade, William Olbers (1758–1840). Olbers actually devised the method for his own use and did not publish it for many years. It never occurred to him that he had come up with a truly novel approach to the problem. After publication, the method became widely used to calculate parabolic orbits and is in use with minor modifications even today. Olbers made numerous other contributions to astronomy, including his famous cosmological paradox that wonders why the night sky is dark. He also discovered Pallas, the second minor planet, in 1802. He is truly one of the great amateur astronomers.

Carl Friedrich Gauss (1777–1865), one of the great mathematical minds, solved the problem of finding the six elements of an elliptical orbit given three observations of the body's position on the sky. He published the result in 1809. Subsequently, techniques have been developed for finding improved orbital elements from a series of many observations. Many of these methods are still in use today.

Gauss and two other greats in the field, Joseph L. Lagrange (1736–1813) and Pierre Simon Laplace (1749–1827), helped devise a mathematical system of celestial mechanics that was used to calculate the motion of bodies under the influence of planetary perturbations. These techniques were especially valuable in days before computers, when the almost endless calculations were executed laboriously by hand. Today, the approach is different. One begins with the basic physical equations "à la Newton," forgets about the sophisticated mathematical system of the nineteenth century, and solves the basic equations using powerful electronic computers. Raw computer power has replaced the older methods very effectively.

COMET STATISTICS

Brian Marsden of the Harvard–Smithsonian Astrophysical Observatory in Cambridge, Massachusetts, is the author of several editions of the *Catalogue of Cometary Orbits,* published by the observatory. The fourth edition (1983) of the catalogue lists orbital elements for 1109 appearances of comets. We say "1109 appearances," because 121 of the comets are periodic comets that have been seen more than once. (A surprisingly large number of periodic comets have been seen only once; they have then been lost for any of a number of reasons.) The long-period comets—comets with period over 200 years—account for 589 entries in the catalogue. The long-period comet with the shortest period is Comet 1905 III with a period of 226 years. Astronomers now consider short-period comets to be those with periods less than about 30 years, and define an intermediate-period group with periods between about 30 years and 200 years. In the 1983 *Catalog of Cometary Orbits* there is a gap between the 37.7-year period of Comet Stephan–Oterma and the 57.5-year period of Comet Pons–Gambart. If we choose this gap as the dividing line between short-period and intermediate-period comets, then there are only 14 known intermediate-period comets. Incidentally, Comet Herschel–Rigollet is the longest period comet (period 154.9 years) to be seen more than once; it was Comet 1788 II and 1939 VI.

Several important points can be determined from Marsden's

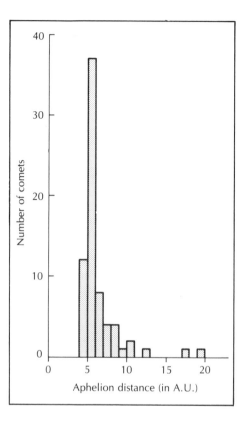

Figure 4–10. A histogram of the number of short-period comets versus aphelion distance. The large spike at 5 A.U. is clearly shown. (Data from the Catalogue of Cometary Orbits, *4th Edition, B.G. Marsden, Harvard–Smithsonian Astrophysical Observatory, Cambridge, Massachusetts, 1982.)*

Catalogue. First, the short-period comets move more nearly like the planets than the long-period comets. Most of the orbits of short-period comets have small inclinations to the plane of the ecliptic and they move in the same direction (direct *versus* retrograde) as the planets. By contrast, the long-period comets show all possible inclinations, and as many move retrograde as direct in their orbits.

Figure 4–10 is a histogram that shows the number of short-period comets with various aphelion distances. (Aphelion distance is the distance at which the comet is as far as possible from the sun in its orbit.) Notice the big spike at 5 astronomical units,* which is Jupiter's distance from the sun. Jupiter has captured a family of comets, and its strong gravitational influence dominates their motion.

---

*The astronomical unit, abbreviated A.U., is the average distance between the earth and the sun. The A.U. is roughly 93,000,000 miles or 150,000,000 kilometers.

CHAPTER 5

WHERE DO COMETS COME FROM?

Our discussion so far has shown how modern ideas have developed up to the beginning of the twentieth century. From the perspective of 1984 then, what are the "vital statistics" of a comet? And where do they originate? We have the benefit of having observed several bright comets in great detail from the earth's orbit in the last decade, with particular attention paid to Comet Kohoutek. However, we have not yet had the benefit of sending a space probe to observe a comet at close range. We stand at a crossroads. Cometary science will continue to advance without a space probe, but at a slow pace. There may be some questions that will never be answered from the earth or its near environs, particularly those about the details of the nucleus. A space probe might radically modify our view of cometary physics, or it might just fill in some vitally needed detail. Either way, a probe would quicken the pace of the field. What we want to describe here and now is our view from the crossroads.

Comets are born, live out a life of variable length, and die. To discuss the birth of a comet, we must look at the birth of the solar system, because the two events are intimately tied together. The theory for the origin of the solar system that is most widely accepted today is an outgrowth of the nebular theory proposed by Immanuel Kant in the eighteenth century and expanded by C.F. von Weizsäcker, G.P. Kuiper, and others between 1940 and the present. There is one concept that we must bear in mind as we discuss the origin of comets, and that is the Oort Cloud.

THE OORT CLOUD

When the orbits of long-period comets are examined when they are still far outside the region of the major planets, it is clear that the points

Figure 5–1. Jan H. Oort.

of greatest distance from the sun—the so-called aphelion distances—lie between 50,000 and 150,000 A.U. This is not a simple analysis to make, however. Since comets are usually discovered when they are in the vicinity of the sun, their motion has already been perturbed by the action of the major planets. As a result, astronomers must work backward and correct the observed motion for these perturbations. This can be done quite precisely today. The Dutch astronomer J. H. Oort (Figure 5–1) suggested that these long-period comets dropped out of a giant cloud of comets that lies between 50,000 and 150,000 A.U. from the sun. This cloud could represent a region of the solar system where densities of particles were sufficiently low that collisions were not frequent enough to grow planet-sized bodies. The outer edge of the cloud at 150,000 A.U. or so is halfway to the nearest star.

Stars in the local region of the galaxy behave somewhat like molecules in a low-density gas, moving about with random velocities that average around 30 km/sec. A relatively simple calculation shows that every 10 million years or so a star of average mass 1.4 times the sun's mass will pass through this cloud of comets and reach a distance of about 50,000 A.U. from the sun. Such a star will have a major gravitational effect on a large number of the comets out in the cloud. Some of them will be torn out of the cloud and will leave the solar system never to return. However, others will be dropped into the inner solar system to become the observed long-period comets. To account for the observed number of long-period comets, the cloud must contain a total of about 10^{11} to 10^{12} comets, which adds up to a mass of approximately a few percent of the mass of the earth. This hypothetical cloud of comets, which is consistent with all of our observations, is likely to be the current configuration of most comets in the solar system. Any theory of the origin of comets should therefore explain the existence of the cloud.

ORIGIN OF THE SOLAR SYSTEM

Our solar system was formed roughly five billion years ago, out of a dense cloud of interstellar material (Figure 5–2). Astronomers believe that some event compressed the primordial cloud to a critical point where the mutual gravitational pull of parts of the cloud on one another caused it to begin to contract. There are at least two known processes that could cause the initial compression, and the presence of some unusual atomic isotopes in the modern solar system offers one hint as to which one it might have been.

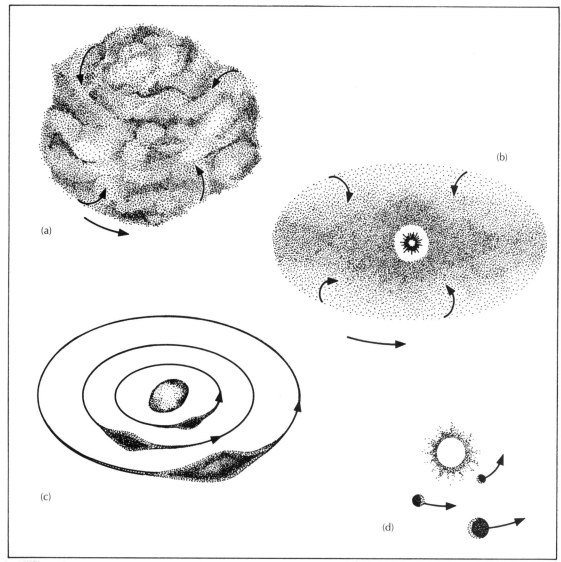

Figure 5–2. The formation of the solar system. (a) The primordial solar nebula or cloud is rotating slowly and the mutual gravitational pull of the parts of the cloud causes the cloud to contract. (b) The contraction of the cloud heats the innermost portion, which becomes the sun. (c) Blobs of material are left in orbits around the sun and they coalesce to form the protoplanets. Heat from the sun evaporates any icy material and the solar wind carries any leftover material out of the solar system. (d) These processes are complete and the solar system reaches its present state. (After New Horizons in Astronomy, 2nd ed., J.C. Brandt and S.P. Maran, W.H. Freeman and Co., San Francisco, California, 1979.)

Figure 5–3. Spiral galaxy in the constellation Andromeda. This galaxy is similar to our own Milky Way galaxy (Palomar Observatory photograph).

We live on the outskirts of a spiral galaxy (Figure 5–3), a body so immense that it takes 100,000 years for light traveling at 300,000 kilometers each second to cross from one side to the other. The spiral arms are actually spiral-shaped waves—not unlike sound waves—that travel through the rarefied gas between the stars. When the wave passes a point in space, it compresses the local interstellar gas. There is strong circumstantial evidence that the passage of a spiral wave must compress some interstellar clouds beyond the critical density for star formation. Most of the young, hot stars in our galaxy lie just inside a spiral arm, along a line that the spiral wave has just passed. These young, hot stars are energy gluttons and burn themselves out in a few

Figure 5–4. The Crab Nebula, a remnant of a supernova explosion observed in 1054 A.D. (Lick Observatory photograph).

million years. Before the spiral wave has moved very far, they are dead and gone. Cooler stars like the sun presumably are also formed in clouds that were compressed by the passage of a spiral wave. However, they are seen in places other than just inside spiral arms because they are so very long-lived that they continue unchanged for several rotations of the waves around the galaxy. Perhaps our solar system was formed because of the passage of a spiral density wave. However, the observation of unusual atomic isotopes, described below, hints of an alternative cause.

A supernova is the death throe of a massive star (Figure 5–4). As a result of changing nuclear reactions at the center of a star, the core of the star suddenly collapses and blows the star apart in an unimaginably huge explosion. The result is a blast wave that expands into interstellar space, compressing the gas and contaminating it with radioactive fallout from the explosion. If a star and its retinue of planets and other lesser bodies is formed from the compressed material, some of the fallout material will remain in these bodies.

Scientists have discovered unusual isotopic anomalies in detailed studies of material from meteorites. They have found evidence in the isotopes of oxygen, neon, magnesium, calcium, krypton, strontium, xenon, barium, neodymium, samarium, and titanium that the material had been involved in intense nuclear processes in the past. They believe that these isotopic anomalies resulted when freshly-synthesized material from a supernova was injected into the primordial solar cloud. The supernova either caused the initial collapse of the cloud, or it exploded soon after the passage of a density wave when the system was just forming.

Once the interstellar cloud began its collapse, a series of processes occurred that led to the various components of the solar system. One fact to bear in mind is that interstellar material is dusty, as has been determined by the effect of intervening dust on the light from distant stars. As the cloud shrank and its density increased, fluffy layers of ice froze onto the dust, giving the particles the appearance of loose snowballs. When these snowballs collided, they stuck together, and natural collisions tended to build up increasingly large solid particles.

In the meantime, the central part of the cloud collapsed faster than the outer portions, building up a protosun. Eventually, the protosun heated up enough to evaporate some of the ices from the solid particles nearby. By a slow process of coalescing, the solid particles built up into larger and larger bodies. The bodies in the warmer, inner solar system ultimately become the solid planets Mercury, Venus, Earth, and Mars. Here the icy component boiled away because of the heat of the forming sun. The bodies in the colder, outer solar system became the gaseous/icy planets Jupiter, Saturn, Uranus, and Neptune.

Not all the bodies throughout the solar system developed into planets. In the inner solar system, the smaller bodies (except for the asteroids) either collided with the planets, leading to the intense cratering now observed, or gradually diffused outward due to gravitational perturbations by the planets. In the outer solar system, where the density was presumably lower, collisions were fewer and many smaller bodies remained undisturbed. Therefore, out around Uranus and Neptune kilometer-sized snowballs remained, filled with interstellar dust. The snow itself is composed of the condensable frozen gases in the cloud from which the solar system formed. These bodies, far from the sun, are comets in storage—the dusty snowballs are the nuclei of comets to be. An argument for the formation of the comets out around Uranus and Neptune is provided by the observed composition of nuclear ices—which appear to be mostly water. Scientists understand how condensation processes take place at various temperatures, and the observed compositions of cometary nuclei are consistent with formation at temperatures of around $-170°C$ (100 K). The composition would be significantly different if condensation took place at a temperature either 50°C warmer or 50°C cooler. At the lower temperature, for instance, water ice will not form. Instead, the water will be tied up in solid hydrates of methane and ammonia.

According to a model proposed in the 1970s by A.G.W. Cameron of the Harvard–Smithsonian Center for Astrophysics, the $-170°C$ temperature occurs in the vicinity of the present outer planets, that is, in the vicinity of Uranus and Neptune. The temperatures out at the distance of the Oort cloud would be far too low for ices with the proper composition to form. Is the picture we have given for the formation of comets incompatible with the Oort cloud concept? Although there are major uncertainties, the answer appears to be no, if we consider gravitational perturbations by Uranus and Neptune.

In 1982, Paul R. Weissman at the NASA/Jet Propulsion Laboratory attempted to carry out computer simulations of the origin and history of the Oort cloud. The models show that if, on the one hand, comets formed near the outer planets, then only about 15% of the original mass of comets now exists in the Oort cloud. The process of gravitational perturbations by Uranus and Neptune that moves the comets out to the cloud is not very efficient, and many comets are swept up by Uranus and Neptune. If, on the other hand, comets formed much farther out in the system, as much as 70% of the original mass could have made it into the Oort cloud. However, then the problem that must be addressed is that the temperatures far from the primordial sun are too low for icy bodies like observed comets to form. Oort also pointed out that the density of material in the primordial solar nebula at the currently suggested distance of the hypothetical comet cloud was far too low for comet formation to occur. In the face of the problems of density and temperature of formation, we should probably lean toward the idea that many comets (up to 10^{13}) were formed near Uranus and Neptune, and that most were subsequently swept up by these planets or lost to the solar system. A few of the original comets were suitably perturbed by Uranus and Neptune and finally wound up in the Oort cloud.

The mechanism that we have outlined above for the formation of the Oort cloud is nearly that suggested by Oort himself in 1950. We are not on very firm scientific ground, though. There is some evidence

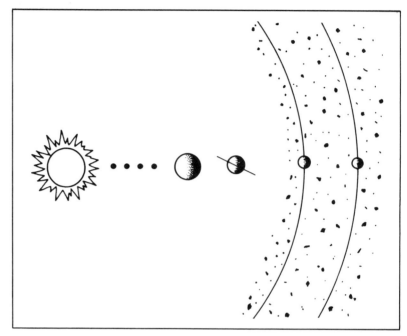

Figure 5–5. Our best current idea for the formation of comets. The physical conditions necessary for formation are found in the solar nebula at the distances of Uranus and Neptune. The dots represent newly formed comets. Formation is the natural result of the broad scheme for the origin of the solar system outlined in Figure 5–2 and discussed in the text. The comets would be ejected into the Oort cloud by gravitational interaction with Uranus and Neptune. This theory is basically the same as that proposed by J. H. Oort in 1950. (Drawing after F. Reddy.)

that the true mechanism for the cloud's origin may never be identified unambiguously. Paul R. Weissman points out that the processes that have randomly distributed comets throughout the cloud have left little evidence of the cloud's origin. A number of different guesses as to the original configuration of comets early in the solar system all lead to essentially the same result.

The scenario presented here is illustrated in Figure 5–5. Of course, we are not certain of many aspects, and some alternative scenarios are presented in Figure 5–6.

In the study of dense interstellar gas clouds, especially by the use of microwave radio techniques, a long list of molecules in the clouds can be made, including some complex organic molecules. The snows making up pristine cometary nuclei probably contain these molecules

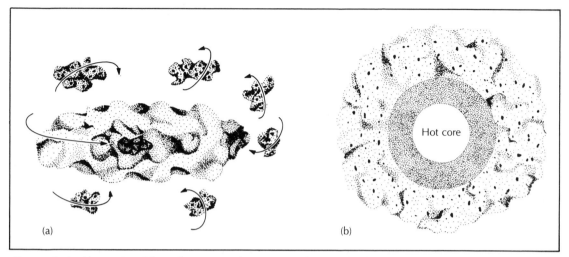

Figure 5–6. Alternative ideas for comet formation. The newly formed comets are represented by dots. (a) The appropriate physical conditions for formation might be found in fragments of the solar nebula in elliptical orbits around the inner solar system. This idea was suggested by A.G.W. Cameron. (b) The physical conditions needed might be present in the outer parts of the protosun itself, as suggested by J. Hills. (Drawings after F. Reddy.)

in frozen form. There is, no doubt, much information about the primordial gas cloud frozen into cometary nuclei.

INTERSTELLAR COMETS

If some comets are torn out of the Oort cloud and leave the solar system, then some day they may encounter another star. The converse must also be true. All stars probably form by more or less the same processes as the sun, in which case many stars probably have clouds of comets surrounding them. It is not out of the question that a few comets escaping from other stars might reach the solar system. These comets would be expected to move in hyperbolic orbits. To date, no comet with a truly hyperbolic orbit has been observed. Nonetheless, such encounters may be very rare. Some day one may show up. The discovery and study of a truly interstellar comet would be of immense importance.

A MINORITY VIEW

There are a number of other ideas about the origin of comets that place the site in the realm of the giant planets and the time in the recent past, astronomically speaking. One idea is that comets are ejecta from super-volcanoes on the outer planets. This is certainly consistent with the composition of comets: the major planets and many of their satellites do have icy interiors. However, the tremendous gravitational fields of the major planets would require the comets to be accelerated to great speeds before they could escape.

In the 1960s, the Russian astronomer Vsekhsvyatskii suggested that volcanic activity in the satellites of the major planets could be responsible for comet formation. The *Voyager* spacecraft found volcanic activity on Jupiter's moon Io (Figure 5–7). The volcanoes on Io have a very high sulphur content. Until recently this was taken as a good argument against Vsekhsvyatskii's hypothesis since comets have not shown any sulphur concentration. However, recent spectroscopic observations of Comet IRAS–Araki–Alcock have revealed the weak presence of diatomic sulphur, S_2. Thus, it is not possible to rule out an origin from Io on these grounds. However, there is growing evidence that Io has virtually no ice, a strong argument against considering it as

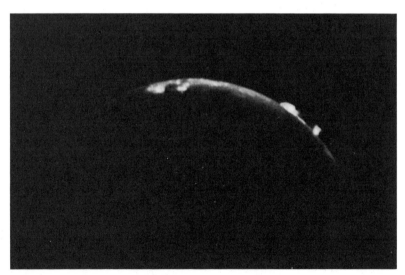

Figure 5–7. Voyager photograph showing volcanoes at the limb of Jupiter's satellite, Io (National Aeronautics and Space Administration, Washington, D.C.).

Plate 1 (above). Comet West as photographed in New England at 4 A.M. on March 7, 1976, by Betty Milon. For observers well away from city lights, the comet was a very impressive sight. The plasma tail (right) shows a blue color while the dust tail (left) appears white.

Plate 2 (right). Possibly prehistoric, this rock art record of a comet was found near Los Alamos, New Mexico (Laboratory for Astronomy and Solar Physics, NASA/Goddard Space Flight Center, Greenbelt, Maryland).

ISTI MIRANT STELLA

HAROLD

MALUM ASTRUM HAROLDO PRAEFIGURATUM A.D.1066 = PRO STELLAE RESURRECTO A.D.1986 DELPHINE DELSEMME FECIT

Plate 3. A facsimile reproduction of the portion of the Bayeux Tapestry showing Halley's comet. (Courtesy of Delphine Delsemme.)

Plate 4. Giotto's fresco "The Adoration of the Magi" in the Scrovegni Chapel, Padua, Italy. The fresco was probably completed in 1304 A.D. Halley's comet had appeared in 1301 and may have served as the model for the star of Bethlehem. (Scala/Art Resource.)

Plate 5. Artist's conception of the eroded surface of a cometary nucleus (© William K. Hartmann).

Plate 6. Artist's conception of a cometary nucleus splitting into a number of pieces (© William K. Hartmann).

Plate 7. Comet Ikeya–Seki as photographed on October 29, 1965. The bright yellow color is produced by the strong sodium emission shown in the spectrum, Figure 6–15. (Courtesy of C.R. Lynds, Kitt Peak National Observatory, Tucson, Arizona.)

Plate 8. Comet Humason in 1961, as photographed with the 48-inch Schmidt telescope at the Palomar Observatory. © California Institute of Technology, 1965.

a parent for comets. In addition, the volcanoes seen on Io produce a fine spray, not comet-sized chunks. No signs of volcanic activity occur on the other satellites observed by *Voyager*.

Another possible site for comet formation could be the rings of Saturn. A possible ejection mechanism could be a close encounter between two or more chunks of ice in the rings. From a dynamical point of view this is not impossible; however, the encounter would have to be so close that the bodies would probably merely collide. All in all, a close quantitative analysis of any of these ejection ideas makes them seem highly unlikely.

Another idea, also fairly easily disposed of, is that comets arise from an exploding planet. Strong evidence has been raised against the theory, but it is sufficiently interesting that it merits discussion. The theory began with calculations of the long-term dynamical evolution of the solar system carried out in the 1970s by the Canadian astronomer M. W. Ovenden. He demonstrated that the present arrangement of the planets might be consistent with there having been a Saturn-sized planet in the asteroid belt up to about 10,000,000 years ago. Then that planet, which he has called Krypton, supposedly blew itself apart somehow. T.C. Van Flandern, of the U.S. Naval Observatory, picked up on Ovenden's concept, using the idea that Krypton should have been an icy planet like the other outer planets, and suggested that comets are the debris of the disrupted planet. The fact that the long-period comets we see today have orbits with major axes around 100,000 A.U., and therefore periods around 10,000,000 years, means they could have originated in the explosion of Krypton, could have made one trip to the outskirts of the solar system but remained under the sun's gravitational influence, and are just now returning to the vicinity of their birth.

The Van Flandern idea is supported by several lines of evidence. One is the fact that significantly more than half of the long-period comets travel in a direction opposite to the direction of motion of the planets. This could be explained by the fact that fragments moving in the same direction as Krypton would have that planet's orbital motion added to their motion from the disruption, giving them an added boost so that many of these fragments would have enough energy to escape from the solar system. Fragments moving retrograde would have Krypton's orbital motion subtracted from their motion and fewer would have enough energy to escape. (This direction-of-motion effect could also arise from other factors involved in a flattened rotating system.)

One major problem with the Krypton theory is obvious. What process can totally destroy a body the size of Saturn? Even all-out nuclear war on Earth would merely render the planet uninhabitable.

All the tens of thousands of warheads stockpiled by the world's nations could not blow the earth apart. Perhaps another retrograde planet existed somewhere in the vicinity of Krypton and the two bodies collided. Even then, difficulties with the idea remain. Another major argument against the idea is a bit more complex. Van Flandern claims that there was an ancient coincidence of comet and asteroid positions all near one point. Many scientists feel this is merely an overinterpretation of poor orbit statistics.

Novel ideas such as this deserve to be discussed. However, the results of the discussion so far have not caused the scientific community to give up the Oort cloud concept, a concept that is almost universally accepted.

COMETS AND THE ORIGIN OF LIFE

Fred Hoyle, in England, and his colleague N.C. Wickramasinghe have based a novel idea on the concept of interstellar comets. They argue that the complex organic molecules frozen into cometary nuclei from interstellar space grow into more complex molecules in the ices, eventually forming rudimentary cell-like structures. These structures are spread to planets throughout the universe by encounters with comets, providing the seeds of life. A. Delsemme of the University of Toledo argues that comets brought organic material to the earth four billion years ago (Figure 5–8). This idea is controversial, but not too far out in left field. However, Hoyle and Wickramasinghe have proposed an even more controversial idea.

Outbreaks of "flu" in schools have the puzzling property that they break out almost simultaneously over large geographical distances. If the outbreaks were caused by the spread of a virus overland, then the flu should start in some location and spread much more slowly. Hoyle and Wickramasinghe blame the flu on viruses from above. The organisms carried by comets drift down to Earth, settle over large areas simultaneously, and cause sickness. Crazy? Yes! But it might explain the nature of flu outbreaks. Could the plagues of the Middle Ages have had a similar origin? Serious biologists do not agree

Figure 5–8 (opposite). Popular article discussing the possible influence of comets on terrestrial life. (Copyright © 1982 by Newsweek, Inc. All rights reserved. Reprinted by permission.)

SCIENCE

A Life-Giving Comet?

Probing the origins of life on earth, a biologist and an astronomer have performed the improbable feat of reinventing religion. Conventional science has invoked the workings of chemistry over almost limitless time to bring the order of life out of the planet's primitive chaos. But life seems to have begun rather quickly: the more scientists have looked, the further back they have found signs of life; the earliest fossil cells, about 3.6 billion years old, are almost as old as the solar system itself. Pondering such mysteries, Nobel Prize-winning biologist Francis Crick and Sir Fred Hoyle, the distinguished astronomer, have independently

strata laid down about the time of the extinction show extraordinarily high concentrations of the element iridium. Iridium is very rare on earth but quite abundant in asteroids. The concentrations, Alvarez calculates, point to an asteroid 10 kilometers across, much bigger than any that has landed in historical time. The dust thrown up by such an impact might have blocked the sun long enough to kill marine plankton, cutting off the ocean's food chain at its source. The shading effect of the dust might also have lowered the earth's temperature dramatically, killing large animals such as the dinosaurs. Alternately, in a scenario by geologist

cules of life; in his new book, "Life Itself," he points out that, with minor exceptions, the genetic code is the same for all living creatures that have been tested, suggesting a single source for all life on earth.

Hoyle's thinking on the origins of life has gone through several stages. In a recent lecture, he outlined a theory that was not unlike Crick's. But in his 1978 book, "Lifecloud" (written with Chandra Wickramasinghe), he suggested that primitive living cells originated in comets and were "seeded" on earth early in its history. In "Lifecloud" he also pointed out that earthly organisms are strangely out of tune with conditions in the rest of our solar system; the wavelengths of light that chlorophyll uses most efficiently, for example, are not those in which the sun's spectrum is concentrated. Such speculation outside the mainstream of science has led Hoyle to exactly the view that seemed self-evident in the Middle Ages: that life did not arise spontaneously on earth. According to this theory, the origins of life are inherently unknowable, or at best a problem for the scientists far out in space where it did arise.

JERRY ADLER with JOHN CAREY

How Puberty May Be Tied to Intelligence

The physical and emotional changes of puberty are well documented; now researchers led by Stanford University psychologist J. Merrill Carlsmith are turning up evidence that the timing of sexual maturity may affect intellectual development as well. The group examined data on more than 6,000 adolescent boys and girls, 12 to 18. They report that boys who reached sexual maturity early scored better on standard intelligence and achievement tests than did late maturers. The scientists did not find this general difference among the girls they studied, but they did note one intriguing phenomenon. The girls who matured late matched or outscored their male counterparts in mathematics. Usually adolescent girls begin to fall behind boys in mathematical ability and, in fact, the early-maturing girls did.

The Stanford scientists are still seeking reasons for the connections between maturity and intelligence, but some researchers suggest it has to do with the "lateralization" of the brain. As the brain develops, the right and left sides carry out their functions in increasing isolation. This lateralization seems to be accompanied by greater mathematical—but poorer verbal—aptitude, perhaps explaining why boys, who mature an average of two years later than girls, do better at math and worse at language. The Stanford results suggest that girls who mature late have more lateralized brains, and hence greater potential math ability, than their early-maturing sisters. By the same reasoning, early-maturing boys have more verbal skills than late-maturing ones.

Bettmann Archive

Nineteenth-century vision of a deadly comet: 'Seeding' the young earth

supposed a *deus ex galaxia* to explain the sudden appearance of life on earth: the "seeding" of space by intelligent beings from distant corners of the universe.

Crick and Hoyle may have the most far-out hypothesis, but they are not alone in asking whether life on earth was made possible—or at least influenced—by objects from the far reaches of the solar system. Astrophysicist Armand Delsemme of the University of Toledo believes that the stuff of living things—including hydrogen, carbon and oxygen—came from comets, which brought gas and organic material to a lifeless, airless earth 4 billion years ago. And Berkeley geologist Walter Alvarez has suggested that an enormous asteroid may have wiped out all the dinosaurs, and numerous lesser animals, in the Cretaceous extinction 65 million years ago.

Geologic evidence for such an asteroid is quite strong. At 26 sites around the globe,

Cesare Emiliani, an asteroid might have struck the ocean and vaporized so much water that a greenhouse effect *raised* the earth's temperature. This theory has a certain advantage: since the ocean floor is continuously destroyed and created anew, the crater from such a shock might no longer exist, which would account for the embarrassing fact that it has not been found.

Colonies: If a crater is ever discovered, it may help settle the issue, and there is hope that more precise geological data will illuminate the Alvarez hypothesis. But it is hard to imagine any experiments that could put Crick's and Hoyle's ideas to the test. How, for instance, can one investigate Crick's hypothesis that a higher civilization in another solar system, fearing extinction as its sun began to die, colonized earth with spaceships filled with frozen bacteria? Crick's interest in what he calls "directed panspermia" rests on peculiarities in mole-

with Hoyle, and this idea should be taken with a grain of salt at the present time.

How, then, is a comet born? That question cannot be answered with finality at the present time. However, comets do seem to be an inevitable product of the birth of the planetary system, remnants of the debris left over from the formation of the planets. For this reason alone comets are extraordinarily interesting. A comet freshly ejected from the Oort cloud might be pristine material from the birth of the solar system, and could carry new clues about that beginning.

After birth comes life. What are the major events in the life of a comet? We will look at these events in the next two chapters.

CHAPTER 6

OBSERVING COMETS

The main component of a comet is its *nucleus,* composed of various ices with a smattering of rocky particles. As the nucleus gets near the sun, solar energy causes the ices to sublime, releasing the rocky particles and forming a dusty halo of gas around the nucleus. Energetic photons from the sun tear the molecules in the halo apart, producing less complex molecules and radicals, some of which are neutral and some of which are electrically charged. The neutral molecules form a glowing cloud around the nucleus, which we call the *coma.* The dust and charged particles, called ions, interact with solar radiation and the solar wind and are blown away from the head to form the long *tail.* Finally, some molecules—probably mainly water—are torn apart in such a way that H (hydrogen) and OH (hydroxyl) radicals ultimately remain, forming a huge *hydrogen–hydroxyl cloud* around the comet.

There are many details left out of this thumbnail sketch, and we must look at each component more closely. However, before we begin looking at cometary phenomena, it will be worthwhile to describe the modern observatory, and the tools and techniques the astronomer uses to study comets.

In the following discussion, a number of chemical terms are introduced. The basic building blocks of all substances are the chemical elements, such as hydrogen (H), oxygen (O), sulfur (S), sodium (Na), chlorine (Cl), and many others. These elements can combine into a large variety of chemical compounds such as table salt (NaCl), sulphuric acid (H_2SO_4) or water (H_2O or HOH), which are made up of molecules, that is, of stable combinations of atoms. In chemical reactions between compounds, molecules consistently break up into pieces called *radicals.* For instance, in chemical reactions involving sulphuric acid, the H_2SO_4 molecule breaks up into hydrogen, H_2, and the sulfate radical, SO_4; and water can break up into hydrogen, H, and the hydroxyl radical, OH. Similarly, in comets the complex molecules break up into relatively stable radicals. If an atom, radical, or molecule carries a net electrical charge, it is called an *ion.* In what follows we will encounter the positively-charged molecular ion CO^+ on numerous occasions.

THE OBSERVATORY

The modern observatory is perched high on a mountain top (Figure 6–1), as far from the polluting influences of civilization and as high above most clouds and water vapor as possible. As population centers spread far and wide, it becomes increasingly difficult to find excellent sites for observatories. Nevertheless, a few sites of high quality do still exist.

Figure 6–1. The Joint Observatory for Cometary Research (JOCR) located on a mountaintop near Socorro, New Mexico. One telescope is a wide-angle instrument designed especially for comet work; the other is a general-purpose telescope. (Laboratory for Astronomy and Solar Physics, NASA/Goddard Space Flight Center and New Mexico Institute of Mining and Technology.)

The comet observer practices his trade at any time of the night. However, the most likely times are just after sunset or just before dawn, when bright comets near the sun are visible. The observatory is a special place. The atronomer is often alone in a dark building that is open to the elements. The observatory roof is rolled away or a dome slit is opened for an unobstructed view of the sky. Since heated air rising from the building would have a blurring effect on seeing, the observatory is kept at the same temperature as the air outside. That can be quite pleasant on a balmy summer night, but on a cold winter night it can make the astronomer dream of a warm bed. Heated flight suits developed for combat pilots have helped a lot.

What does the astronomer do in the observatory? Observations can be simply divided into three different types: imaging, photometry, and spectroscopy.

Astronomical imaging can be as simple as taking a photograph of all or part of a comet. Cameras with wide-angle lenses are generally used to obtain portraits of entire comets (Figure 6–2). Large, long-focus telescopes are used to probe the details of the nucleus or other fine details. When the astronomer wants more sophisticated images, a number of tools are available. Special filters can be used to isolate selected emissions of light from the comet. Polarized filters can be used to study the polarization of the light from the comet. The photographic plate can be replaced by any of a number of modern electronic imaging detectors. Imaging techniques have come a long way since the first simple photographs of the sky were obtained in the late nineteenth century.

Photometry means the measurement of the brightness of light. Cometary photometry can be the measurement of the brightness of an entire comet, or of some small part. Sensitive photoelectric devices and specialized filters are used. Photometry got its start when astronomers compared the brightness of a comet to out-of-focus star images in a telescope. Modern photoelectric photometry is similar in concept to a photographer's use of a light meter to measure the light intensity of a subject (Figure 6–3).

Spectroscopy needs a bit more explanation. It is based on the fact that information about the atoms and molecules that make up a heavenly body—a star, a planet, or a comet—is contained in the light it emits and reflects. Natural light is a mixture of electromagnetic waves with wavelengths covering a wide range. When light is passed through a spectrograph it is dispersed, that is, the wavelengths are separated. Then, depending on the physical conditions, a pattern of bright lines stands out against a dark background, or a pattern of dark lines stands out against a bright background. From the wavelengths of

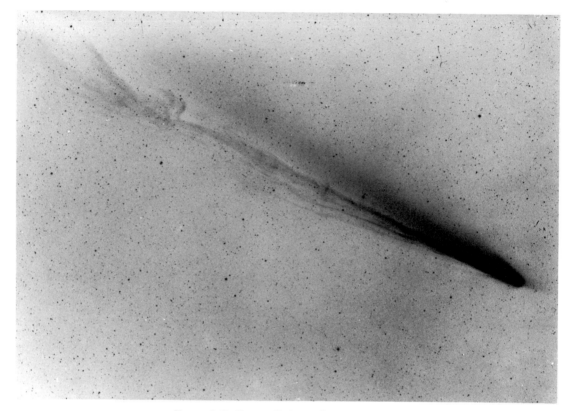

Figure 6–2. Comet Kohoutek on January 11, 1974. This image, made with a wide-angle telescope, shows the "Swan cloud" (upper left) approximately 10 million miles from the head (JOCR photograph).

the lines, we can infer what atoms and ions are present in the light source.

In practicing the art of spectroscopy, the astronomer uses a spectrograph attached to the telescope. The telescope collects the light from a comet and focuses it into the spectrograph. The spectrograph then produces a photographic or photoelectric record of the spectrum (Figure 6–4). Later, the astronomer "reduces" the spectrographic

Figure 6–3 (opposite). (a) Diagram of the photoelectric process. 1. Analogy to a light meter. 2. Method of obtaining observations. 3. The raw observing record. (b) The end result, a graph showing the brightness variation of Comet Kohoutek (1973f) in CN (cyanogen) as a function of distance from the sun. (Data by L. Kohoutek, Hamburg Observatory, West Germany.)

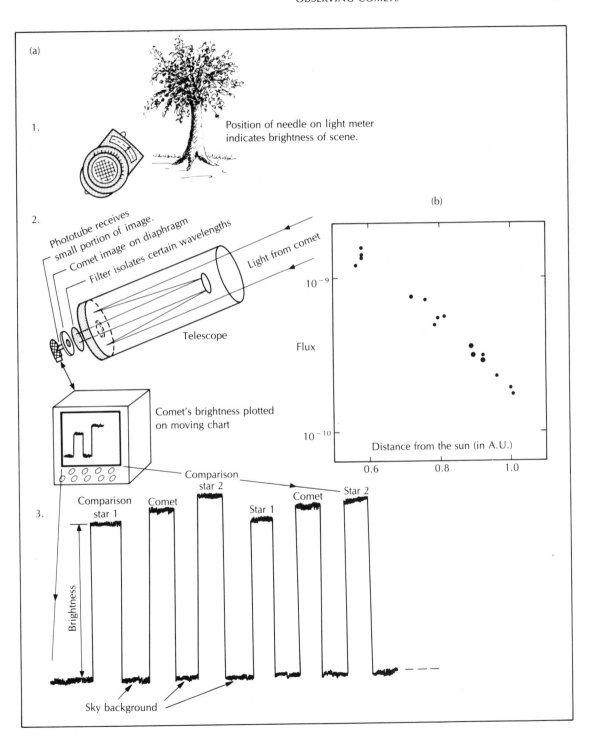

(a)

1. Position of needle on light meter indicates brightness of scene.

2. Phototube receives small portion of image.
   Comet image on diaphragm
   Filter isolates certain wavelengths
   Light from comet
   Telescope

   Comet's brightness plotted on moving chart

3. Brightness

   Comparison star 1
   Comet
   Comparison star 2
   Star 1
   Comet
   Star 2

   Sky background

(b)

Flux

10^{-9}

10^{-10}

Distance from the sun (in A.U.)

0.6   0.8   1.0

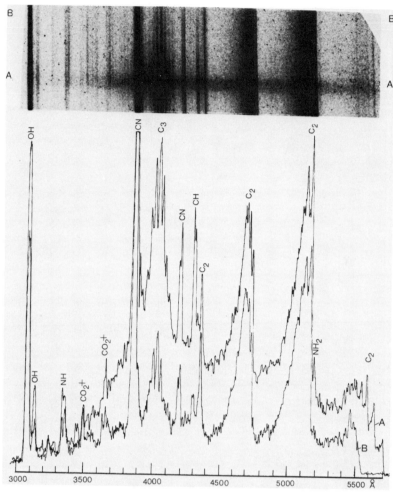

Figure 6–4. The spectrum of Comet Bradfield (1979l) made on February 5, 1980. A is at the nucleus and B is 17,000 km in the tail direction. (Courtesy of S.M. Larson, University of Arizona, Tucson, Arizona.)

record by determining the wavelengths and intensities of each line recorded in the spectrum. Then the astronomer compares those wavelengths with wavelengths of spectra of known substances that have been measured in the laboratory. If a pattern of lines in the cometary spectrum matches a pattern in a laboratory spectrum, then it is possible to conclude that the substance that produced the laboratory spectrum is present in the comet. This technique is satisfactory not only for the visible light spectrum, but also for infrared, ultraviolet, radio, and other wavelength regions. The main differences between the various spectral regions are the techniques that are used to detect and record the spectra.

We said earlier that the experimental side of the scientific method is not directly useful to studies of celestial objects, and that is true. We cannot yet experiment with comets or any other celestial bodies. However, experimentation is important in interpreting observations. Physicists have spent years in the laboratory studying the spectra of terrestrial atoms and molecules. The data that have resulted from their studies are essential to our understanding of cometary spectra.

The first spectroscopic observations of a comet were made in 1864 in Florence, Italy, by Giovanni Donati (1828–1873), who observed Comet Tempel (1864 II). Subsequently, others took up the effort, notably Sir William Huggins (1824–1910) in London. These early spectra showed faint bands of emission and, occasionally, the superimposed spectrum of the sun. The bands of emission observed in the heads of comets were identified in due course as arising from the carbon molecule, C_2, cyanogen, CN, and the molecules C_3 and CH. In some comets, spectrum lines due to sodium were observed as well, as in the spectrum of Comet 1881 III when it was near the sun.

In the early twentieth century, studies of cometary spectra had advanced to the point where the much fainter tail spectra could be analyzed. Interestingly enough, the molecules identified in the tails were all ionized; that is, they had lost an electron and had a net positive electric charge. The primary emissions observed were due to ionized carbon monoxide (CO^+) and molecular nitrogen (N_2^+).

Many more atoms, molecules, and ions have been identified in comets since the early days. We will return to this point. However, let us first take a brief look at the source of all the material: the nucleus.

THE NUCLEUS—FIRST LOOK

If we had a "transporter" that could beam us down to a comet's surface, we would see a truly marvelous place and might file the following report:

The atmosphere has a misty appearance because of what appears to be a heavy concentration of dust. We can see the sun dimly through the haze. (It appears much as it might in Los Angeles on a very smoggy day.) As our eyes adjust to the dim light, we can see many interesting phenomena. There are numerous cracks in the surface and jets of gas shooting up all around. When we dig into the surface we find it is a dusty crust, slightly below which we find fresh ices that immediately begin to sublimate when exposed to the sunlight. The entire atmosphere appears to be blowing away, carrying along the dust that escapes from the nucleus. The cracks and jets appear to be areas of greater than average gaseous emission. The irregularity in gas emission appears to be caused mostly by the loose-packed nature of the dust and ice. The erosion of the surface layers is very irregular as shown in a sketch of the scene (Plate 5). Occasionally, an interplanetary rock strikes the surface, producing a crater and exposing fresh ices, creating yet another jet of gaseous emission.

Our exploration of the comet's nuclear surface reveals additional interesting facts. Walking all the way around the comet, we learn that it is only about three kilometers in circumference, or one kilometer in diameter. By timing the sun as it rises and sets in the mist, we find that the comet's nucleus rotates on its axis every few hours and that the maximum gas emission, on average, occurs in the "afternoon" hours. We also see phenomena much more catastrophic than local fissures and jets. Subsurface hollows occur here and there, and their collapse produces major crust slides. So much fresh ice is exposed by these slides that minivolcanoes are produced. Finally the subsurface rumblings become extremely severe and the entire nucleus begins to split into independent pieces (Plate 6). It is time for us to depart. . .

The cometary nucleus is truly a "magic mountain." This remarkable object, even when it is only one kilometer across, can produce a gas cloud over a million kilometers in radius and a visible gaseous tail some ten million kilometers or more in length. We know that the nucleus can produce the gas cloud and long tail, but our description of the surface and its phenomena was entirely fanciful. No one has ever seen the nucleus of a comet. We do not have even one photograph. Our description *might* be accurate, but it is largely conjecture based on circumstantial evidence.

This is the basic dilemma facing scientists studying comets and those wishing to write about them. The source of the phenomena that can actually be seen and studied is the nucleus, which no one has ever seen. Thus one of the most important goals in cometary science is to see, measure, probe, and understand the nucleus. For this reason, two of the direct missions to Halley's comet in March of 1986 (see Chapter

11) have photograhy of the nucleus as a prime objective. Within a few years, authors of comet books may be able to provide a description of the nucleus and the cometary surface based on actual photographs and direct measurements. Ultimately, spacecraft may land on the surfaces of comets. We will then learn about local features, and samples may be returned to Earth for analysis.

Now, let's look at what scientists have inferred about cometary nuclei.

The first question that might be asked is, how big is the nucleus? There are two ways to address this question: direct measurement and inference from measured brightness with appropriate assumptions.

Normally, a comet's nucleus appears to be a star-like point of light when observed through a long-focus, high-power telescope. It just is not possible to resolve any perceptible disk. At best, an upper limit on the size can be suggested: "If the nucleus were as large as such-and-such, a disk would have been seen." In 1927 periodic Comet Pons–Winnecke passed only 6,000,000 kilometers from the earth. At that time, astronomers measured the angular size of the nucleus to be 0."3. That is a tiny angle, very difficult to measure, and it should be viewed only as an upper limit. An object at 6,000,000 kilometers with an angular diameter of 0."3 is at most 5 kilometers in diameter. Thus, the upper limit on the diameter of the nucleus of Comet Pons–Winnecke was 5 kilometers. Upper limits in other cases range from about 5 to 100 kilometers.

Several comets have passed directly between the earth and the sun. On these occasions, astronomers have tried to detect the nuclei silhouetted against the solar disk. The Great Comet of 1882 is a prime example. No trace of a nucleus was observed during the transit, yet the geometry of the transit was such that a nucleus larger than 70 kilometers would have been seen. The most we can say about this comet is that the diameter of its nucleus is probably less than 70 km. Halley's comet transited the sun in 1910 (see Chapter 10). Observations made at that time showed that its nucleus must be smaller than 100 kilometers in diameter.

Another class of comet that tells us something about nuclear sizes is the sun grazers. These bodies come uncomfortably close to hitting the sun (see Chapter 8). For instance, Comet 1843 I passed roughly 125,000 km above the sun's surface. That seems like a big distance—more than ten times the earth's radius—until we remember it is only 18% of the sun's radius. If the comet's nucleus were only a few meters across, it would have been completely vaporized by the sun's heat. Nuclei of sun grazers must have diameters at least in the many-meter range.

To see how sizes can be inferred from brightness, imagine the nucleus to be a perfect mirror reflecting sunlight back to the earth. The amount of light a mirror intercepts depends on its area. The larger it is, the more light it reflects. In this simple case, if we know the mirror's distance from us, we can immediately calculate its diameter from its measured brightness. Now, of course, comes the rub. The nucleus is not a perfect mirror; not all the energy that hits it is reflected—some is absorbed and goes to sublime the ices. We can only estimate its reflectivity and the direction of the reflection. Since the nucleus has a rough surface, photons will be reflected in all directions, with one preferred average direction. With reasonable values for reflectivity, we find nuclear sizes in the range 1 to 10 km for long-period comets and 1 km and less for short-period comets. Let's pick a nice conservative 1 kilometer as the average size of a nucleus.

What is the mass of a kilometer-sized iceball? Ices have densities in the range of 1 gram/cubic centimeter. Jupiter's and Saturn's icy moons have densities in that range, for example. The diameter gives a volume, and volume × density = mass. The calculation thus gives 4×10^{15} grams or 4 billion tons. Four billion tons seems like a huge number. To put it in perspective: the U.S. bituminous coal industry mined about 4 billion tons of coal in the years 1969 through 1975.

Not much more can be said about the nucleus without looking further at the remainder of cometary phenomena and relating them to the nucleus. Thus, we will now examine the other phenomena in detail.

THE COMA

Short-exposure photographs of a comet show a more or less spherical cloud of material surrounding the nucleus. As exposure times are increased the apparent size of the cloud increases (Figure 6–5). What we are seeing is a cloud whose brightness decreases with increasing distance from the nucleus, eventually blending into the interplanetary medium. The head of the comet is composed of this cloud, together with the nucleus. Spectroscopy shows the head to be composed of neutral molecules, ionized molecules, and dust. The ionized molecules and some of the dust will eventually be swept into the comet's tail, while the neutral molecules form a glowing cloud of material that we call the coma. The ionized molecules passing through

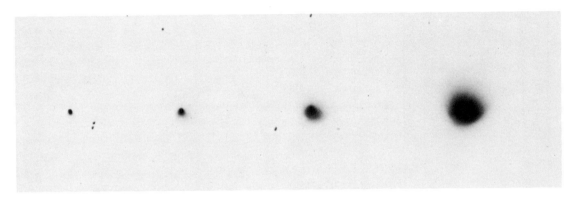

Figure 6–5. Photographs of the coma of Halley's comet on May 25, 1910. The exposure times are (left to right): 15 seconds, 30 seconds, 1 minute, and 3 minutes. (From the Atlas of Cometary Forms, *by J. Rahe, B. Donn, and K. Wurm. NASA SP-198, 1969.)*

the head on their way to becoming tail material are excluded from the definition of the coma.

How big is a typical coma? This question is difficult to answer simply because the coma does not have a sharp edge. The measured size will depend on how the measurement is made. A definition of size might be the distance from the nucleus to the contour where the apparent brightness decreases by a factor of 2, or 10, or some other agreed value. Once we agree upon a definition of the diameter of the coma, we find a second problem. The diameter of the coma of a typical comet changes as its distance from the sun changes. As a comet approaches the sun from deep space, the coma grows in diameter until the comet reaches a heliocentric distance of 1.5 to 2.0 A.U. Then the coma proceeds to shrink as the comet moves even closer to the sun. Coma sizes tend to be between 10^4 and 10^6 kilometers. Thus comae are much larger than cometary nuclei—in fact, they are typically larger than the earth. Figure 6–6 illustrates the varying size of the coma of a typical comet as it approaches the sun from deep space.

Considerable structure has been seen inside the comas of numerous comets, usually by visual observations. Figure 6–7 shows drawings of Comet Tebbutt made in 1861. The presence of streamers originating from the sunward position of the nucleus is obvious in the drawings. The streamers change in appearance quite rapidly. Figure 6–8 is a photograph of Comet Bennett made in 1970. A spiral-shaped pattern of streamers appears near the comet's nucleus. Our best guess as to the origin of the streamers is that the nucleus sublimates more

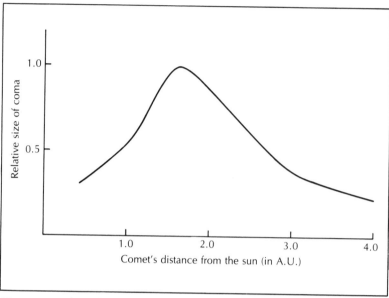

Figure 6–6. The relative size of the coma of an average comet as a function of its distance from the sun. The average comet reaches its maximum coma size when it is 1.65 times the earth's distance from the sun. Other sizes are given in terms of that maximum size. The data were adapted from a statistical study of over 40 comets made at the Hamburg (West Germany) Observatory between 1932 and 1953 by M. Beyer. Accurate measurements of coma size are very difficult and the curve should be considered as indicating a general trend.

rapidly than average at several points on its sunward surface. The spiral pattern observed in Comet Bennett is due to the rotation of the nucleus. The curving shape of the streamers in Comet Tebbutt is due to a repulsive force acting from the direction of the sun (see below).

Many comets undergo sudden outbursts in brightness that may last for up to a month. Structural changes are then observed to proceed even more rapidly than changes in streamers. During the outburst, a shell of material expands slowly away from the nucleus at speeds between 100 and 500 m/s. The expansion continues for nearly a month; then the shell slowly fades away. Periodic Comet Schwassmann–Wachmann I (Figure 6–9) is well known for its outbursts. One idea put forward to explain these particular outbursts is that the nucleus is covered with a powdery layer of soil. Sunlight vaporizes ice until the soil gets so gas-charged that the gas pressure blows it off. Then the comet settles down until it gets recharged and the process

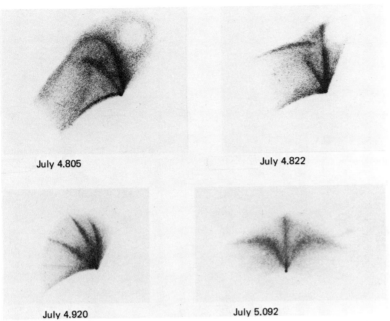

July 4.805

July 4.822

July 4.920

July 5.092

Figure 6–7. Drawings made in 1861 of a spiral structure in Comet Tebbutt. (From the Atlas of Cometary Forms, *by J. Rahe, B. Donn, and K. Wurm. NASA SP-198, 1969.)*

repeats. Interestingly, such outbursts appear to be not at all rare. For instance, outbursts were observed on 59 of the 79 comets known to pass perihelion between 1935 and 1975. The other 20 could easily have produced outbursts, too; they were all unobserved for extended periods due to poor weather conditions or poor placement on the sky.

The spectrum of a comet contains clues to the material that makes up the nucleus, as well as to physical conditions in the cometary gas. Table 6–1 lists the atoms and molecules observed in cometary spectra. Infrared observations tell us that dust grains are present in the head. Telltale bumps in a comet's spectrum at 10 microns and 18 microns indicate the presence of silicate—good old everyday sand.

The typical behavior of the spectrum as a function of a comet's heliocentric distance is a clue to bear in mind. At roughly 3 A.U. from the sun bands of cyanogen (CN) may appear in the spectrum, and at 2 A.U. bands of C_3 and NH_2 may appear. As a comet moves even closer to the sun, C_2, OH, CH, and NH may appear near 1.5 A.U. and grow in strength as heliocentric distance decreases. The first atomic lines, due to sodium (Na), appear at about 0.8 A.U. For those comets that

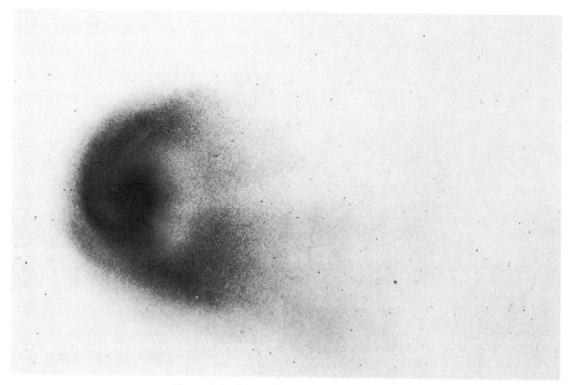

Figure 6–8. Comet Bennett, photographed on March 20, 1970, with spiral structure in the head region. (Courtesy of S.M. Larson, University of Arizona, Tucson, Arizona.)

1961   Oct. 12   Oct. 18   Nov. 3

Figure 6–9. Comet Schwassmann–Wachmann I in 1961 with an outburst clearly shown. (Courtesy of E. Roemer, University of Arizona, Tucson, Arizona; official U.S. Navy photograph.)

Table 6–1. Atoms and Molecules Observed in Cometary Spectra

Atoms		Molecules		Ions
H	Hydrogen	C_2		C^+
C	Carbon	CH		Ca^+
O	Oxygen	CN	Cyanogen	OH^+
Na	Sodium	CO	Carbon monoxide	CH^+
S	Sulfur	CS		CN^+
K	Potassium	C_3		CO^+
Ca	Calcium	HCN	Hydrogen cyanide	N_2^+
V	Vanadium	CH_3CN	Methyl cyanide	H_2O^+
Mn	Manganese	NH		CO_2^+
Fe	Iron	NH_2		
Co	Cobalt	N_2	Nitrogen	
Ni	Nickel	S_2	Sulfur	
Cu	Copper	OH	Hydroxyl	
		H_2O	Water*	
		NH_3	Ammonia	

*Water (H_2O) has been observed weakly at radio wavelengths. Its identification remains questionable.

approach within 0.1 A.U., lines of metals such as nickel (Ni), iron (Fe), and chromium (Cr) may appear in the spectrum.

What we have discussed so far permits us to understand something about the physical conditions in the nucleus and coma. To complete the picture, however, we should discuss the tail.

THE TAIL

Cometary tails point away from the sun, no matter where the comet is situated in its orbit. In the nineteenth century, comet tails were classified according to their shape: type I tails were straight and type II tails were gently curved (Figure 6–10). Some type of force from the direction of the sun was acting on the materials in tails and apparently repelling them. The force was stronger in type I tails and weaker in type II.

Several ideas concerning the nature of this repulsive force have been suggested over the centuries, including electrostatic repulsion.

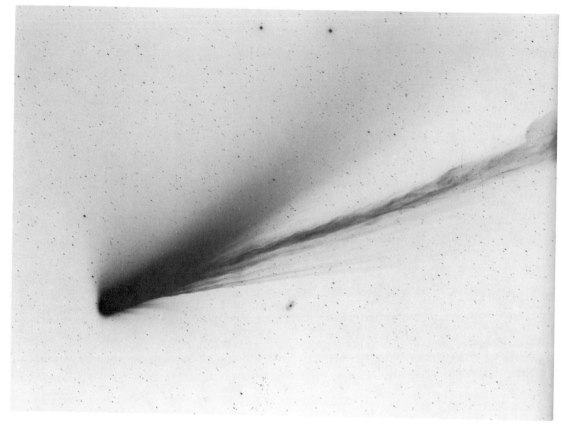

Figure 6–10. Comet Mrkos on August 24, 1957. The upper, diffuse tail is the type II or dust tail. The lower, straight, structured tail is the type I or plasma tail (Palomar Observatory photograph).

The problem with electrostatic force is that the sun would have to have a net charge, and there is a great deal of evidence that the sun is electrically neutral. In the early years of the present century it was suggested that radiation pressure does the job. To understand how this might happen, it is easiest to think of light as a particle phenomenon. A light beam is made up of an immense number of particles called photons. Like moving balls, photons have momentum; when they bounce off a material object they exert a force on it. A cue ball colliding with a stationary billiard ball and sending it into a pocket is a good example of the action of momentum. Atoms and molecules colliding with a container wall exert pressure on the wall. Photons carry tiny amounts of momentum, and so objects of the size we encounter every day never notice their effects. However, a material particle with a diameter of a micron or less will be given a definite shove if hit by a photon. When the spectra of type II tails were obtained, they were found to be merely reflected solar spectra. It is clear that the tail is mostly tiny dust particles. The solar photons reflecting from them push them away from the sun. This physical process, which we call radiation pressure, fits the observations and is now fully accepted as one of the forces that keep comet tails pointing away from the sun.

Type I tails presented a different problem. Spectroscopy showed that in such tails the primary emission is the blue glow of the ionized molecule CO^+. How does radiation pressure move molecules? The original idea was simply based on the fact that all the solar photons come from one direction—the sun. A photon is absorbed by a molecule and gives up its momentum. Then it can be re-emitted in any direction, and the molecule returns some momentum to the photon. The molecule will probably not re-emit the photon in exactly the same direction it absorbed it, and so will retain some of the momentum in the original direction of the photon. On the average, an ensemble of CO^+ molecules will be given a net push away from the sun due to the undirectional absorption and multidirectional re-emission. We will return to this point later (Chapter 7), because a more recent development has significantly changed the explanation of the repulsion in type I tails. Another force is required.

Type II tails are composed of dust, and so we shall refer to these tails as "dust tails" in what follows. Modern calculations have modeled the shapes of dust tails quite well. As the ices in the nucleus sublime, dust particles are released and then accelerated outward by the expanding coma gases. Eventually a point is reached where only two forces act on a dust particle: gravitation and radiation pressure. The tail shape is determined by the ejection speeds, sizes, and ejection rates (which may vary with time) of the dust particles.

Spectroscopy of type I tails has shown the presence of numerous other molecular ions in addition to CO^+, including CO_2^+, H_2O^+, OH^+, CH^+, and N_2^+. The spectra permit us to infer densities in the tails. For instance, the CO^+ ion ranges in density from 1000 ions per cm^3 near the head to 10 ions per cm^3 near the end of the detectable tail. Even though tails can range in length from 10^6 to 10^8 kilometers, very little matter is present at such low densities.

A gas composed of ionized molecules or atoms is called a plasma. Spectroscopy clearly shows that type I tails are composed of plasma. Thus it is better to refer to a "plasma" tail than to a "type I" tail. "Plasma" contains some physical information, while "type I" is merely arbitrary.

THE NUCLEUS—SECOND LOOK

The preceding brief review of the properties of a typical coma and tail is a summary of an immense amount of data gathered from many different comets. A detailed study of all the data has led to a fairly well-accepted model of the cometary nucleus (Figure 6–11).

An early idea theorized that the nucleus was a collection of porous, non-icy, meteoritic solids with large quantities of material adsorbed into the surface. This model was called the sand-bank model. The adsorption–desorption model does produce gases at the proper *rate* to result in all observed cometary phenomena. However, the model must be rejected for a very simple reason. It is just not possible to adsorb enough material onto any known solid to produce a comet with the staying power of, say, Comet Halley. This splendid comet would have outgassed all its adsorbed material and disappeared long ago. In fact, at the rate Halley produces gaseous material, it probably would have disappeared a few weeks after its birth if it *lived* on adsorbed material. The only known way a comet can produce the observed quantity of gaseous material, pass after pass, is for the bulk of the nucleus to sublimate. Thus, the nucleus must be pictured as a large body of ice and snow, with dust particles imbedded in it. The dust is

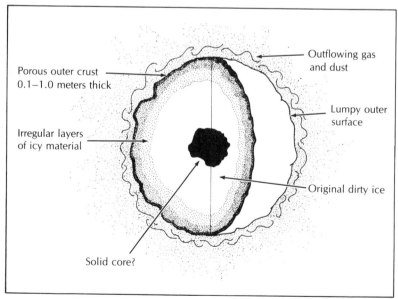

Figure 6–11. Summary model of the comet nucleus.

released as the ice sublimates. This picture was originally proposed in 1950 by F.L. Whipple (Figure 6–12) of the Harvard-Smithsonian Center for Astrophysics, and is called the icy-conglomerate model.

Whipple's icy-conglomerate model agrees with the data very well. However, as new information on the composition of cometary ices has been obtained from spectroscopy and from observations of the physical behavior of the nucleus, the theory has been modified. Earlier, we pointed out that the first bands observed in a cometary spectrum—bands of the cyanogen molecule (CN)—begin to appear in the spectrum when a comet is roughly 3 A.U. from the sun. Thus, the primary constituent must be one that can sublimate at significant rates at the temperatures a comet would experience due to solar heating at 3 A.U. The only known substance that will fill that prescription is water. Thus, astronomers believe the cometary nucleus to be mostly water. This belief is not inconsistent with the fact that CN is the first molecule to be observed in the spectrum. In some way, CN makes up a trace constituent mixed in with the water. It is difficult to observe the water spectrum at wavelengths accessible from the ground, but the CN spectrum is easily observed.

Figure 6–12. Fred L. Whipple.

A.H. Delsemme (Figure 6–13) and P. Swings (in Belgium) have shown that under the conditions found in cometary nuclei, a type of substance known as a clathrate hydrate will be formed. Basically, water molecules attract one another strongly because of a force called the hydrogen bond, first discussed by Nobel Prize winning chemist Linus Pauling. When water freezes into ice, it forms a loose crystal structure, held together by the hydrogen bond, with significant cavities in the structure. This loose structure is the reason why solid ice has a lower density than liquid water; hence ice floats. When water crystallizes in the presence of volatile substances, the cavities in the water crystals are large enough to trap atoms and molecules. Delsemme and Swings showed that if we could greatly magnify the crystals in cometary ices, we would find water ice with its cavities filled with molecules of methane, ammonia, and many other substances. When the ice sublimates, these molecules are released from their icy prison cells. If the picture we presented in Chapter 5 of the origin of comets is correct, then we would expect the molecules trapped in the water ices to include many of the organic molecules now observed in interstellar space: formaldehyde (H_2CO), isocyanic acid (HNCO), cyanoacetylene (HC_3N), formic acid (HCOOH), methyl cyanide (CH_3CN), acetaldehyde (CH_3CHO), and others. Of this list, only methyl cyanide has possibly been detected, as a weak microwave radio emission.

A fundamental problem in the detection of these molecules is the

Figure 6–13. Armand H. Delsemme.

fact that they are quickly torn apart (or dissociated) by the intense ultraviolet radiation from the sun as soon as they are released from the nucleus. Many of the molecules observed in cometary spectra (OH, CH, CO, CO^+, CN) are probably only pieces of more complex molecules that have been dissociated by sunlight. Scientists speak of the parent molecules, which exist within the nucleus, and first-generation molecules, the pieces that remain when the parent molecules are dissociated. One of the big questions in cometary science is, What are the parent molecules? It is possible to infer what they might be, but their existence must be verified somehow.

We have now seen what the nucleus must look like. It is a mass of water ice formed, perhaps, of loosely packed snow crystals with other molecules trapped in the crystal lattice. Imbedded in the mass are solid grains that are released as the ice sublimates. The sublimation process does not proceed in a uniform way; that is, the nucleus does not disappear like an onion being peeled, layer after layer. Instead, it proceeds in a very nonuniform way, leaving large cavities in which quantities of gas may be trapped. The sudden release of gas from such pockets may account for the outbursts observed in so many comets. There may also be an ice-depleted dusty layer or crust on the surface of the nucleus. The process of sublimation must slowly eat away at the nucleus, causing it eventually to disappear. We will explore the ramifications of that statement after we look at an important physical process that occurs in the gaseous/plasma part of a comet.

HYDROGEN–HYDROXYL CLOUD

In 1970 Comet Tago–Sato–Kosaka was observed by the second Orbiting Astronomical Observatory (OAO-2) at an ultraviolet wavelength that shows emission due to hydrogen gas. The observation revealed a huge cloud of hydrogen about 0.1 A.U. in diameter surrounding the comet. Subsequently, similar clouds were observed around Comet Bennett, Comet Encke, and Comet West (Figure 6–14). Further studies of the ultraviolet spectra of these comets show a hydroxyl (OH) cloud, but it is much smaller than the hydrogen cloud. This complex is now called the hydrogen–hydroxyl cloud.

Astronomers believe that the presence of both H and OH is strong supporting evidence for the presence of water, since water (H_2O, or more accurately, HOH) is dissociated into H and OH by ultraviolet photons.

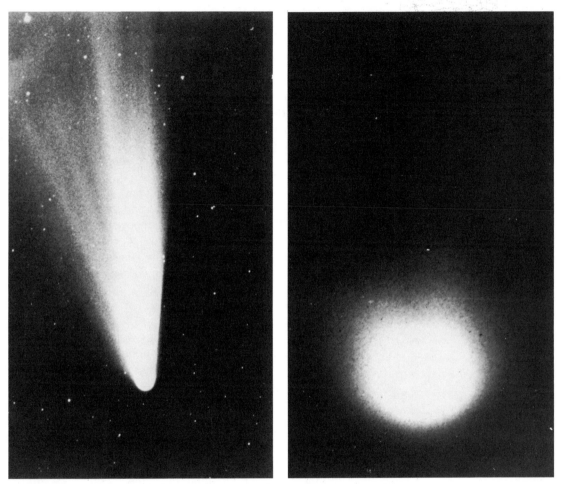

Figure 6–14. Photographs of Comet West taken from a rocket on March 5, 1976. (Left) Visual light image. (Courtesy of P.D. Feldman, Johns Hopkins University, Baltimore, Maryland.) (Right) Ultraviolet image in Lyman-α showing the hydrogen cloud. (Courtesy of C.B. Opal and G.B. Carruthers, Navel Research Laboratory, Washington, D.C.) The two images are at the same scale.

0 1^0

EMISSION LINES

When the spectrum of a comet that is within about 3 A.U. of the sun is examined, emission bands due to many different molecules are seen.

These emission bands occur because the molecules in the comet emit radiation at a series of frequencies characteristic of these molecules. The first question to ask is, Where does the energy emitted by the comet come from? The answer to this question was suggested in the early years of this century by a number of astronomers. It was finally developed on a sound theoretical basis in 1928 by the Dutch astronomer H. Zanstra. Zanstra showed that the emission spectra of comets are caused by an excitation of the molecules by sunlight and subsequent re-emission of the radiation by the cometary molecules. The process is known as fluorescence if the molecule re-emits a longer wavelength (lower frequency) than it absorbed, and as resonance fluorescence if it re-emits the same wavelength it absorbs. Zanstra carried out quantitative calculations for several comets, particularly Comet Wells (1882 I), and showed that the theory predicts the correct brightnesses for comets.

There was one small problem with the Zanstra idea, which can best be explained by the spectrum of CN in comets. The various lines that make up the bands of CN observed around 3880 Å* (for example, Figure 6–4) in cometary spectra had peculiar intensity ratios that changed with the comet's position in its orbit. That is, lines that should have been intense in the CN bands frequently were much too faint. In 1943, the Belgian astronomer P. Swings hit upon the answer. The sun does not emit uniformly at all wavelengths. Many strong absorption lines are present in the solar spectrum. These lines are essentially wavelength regions where the solar energy output is suppressed. Any line in the CN bands with a wavelength corresponding to the wavelength of a solar absorption line will be weak or absent. In one case, two strong absorption lines due to neutral iron in the solar atmosphere (wavelength 3878.02 Å and 3878.57 Å) cause a large decrease in the strengths of two lines of CN. In addition, as the comet moves around its orbit, its velocity relative to that of the sun changes and the Doppler effect shifts the solar spectrum relative to the cometary spectrum. The result is that absorption lines in the solar spectrum appear to move in wavelength with respect to emission lines in the cometary spectrum, and the emission lines that are affected by the absorption lines change accordingly. This effect is not known as the Swings effect, after the man who explained it.

The spectrum of Comet Bradfield (Figure 6–4) is representative of optical wavelength cometary spectra when comets are at heliocentric distances of roughly 1 A.U. While there is major variation from

---

*Ångström. One angstrom = 10^{-8} of a centimeter.

Calcium

3934 Å   3968 Å

5890 Å   5896 Å

Sodium

Figure 6–15. High resolution spectrum of Comet Ikeya–Seki taken on October 20, 1965. At that time, the comet was very close to the sun. Many of the spectral features are characteristic of a solar spectrum reflected from the comet's nucleus and dust, as illustrated by the calcium lines at 3934 Å and 3908 Å. Many of the other lines are due to iron. The emission lines of sodium at 5890 Å and 5896 Å are the most conspicuous features in the spectrum and are produced by sodium atoms from dust grains vaporized by the sun's radiation (Sacramento Peak Observatory, Sunspot, New Mexico).

comet to comet, these spectra are dominated by molecular bands. This situation changes dramatically if the comet passes very near the sun. At heliocentric distances of one-tenth of an astronomical unit or less, the spectrum is dominated by the emission lines of metals superimposed on a reflected solar spectrum, as shown in the spectrum of Comet Ikeya–Seki (Figure 6–15). Most of the conspicuous emission lines are iron. Exceptions are the two very bright lines of sodium (marked). These yellow lines are responsible for the color of Comet Ikeya–Seki as shown in Plate 7.

Orbiting observatories have provided us with the opportunity to make routine observations of comets in the wavelength range 1200 Å to 3200 Å, wavelengths not observable from the ground. Spectra of Comet Bradfield obtained from the *International Ultraviolet Explorer (IUE)* are shown in Figure 6–16. As might be expected, the spectrum is dominated by hydrogen and hydroxyl emission. One characteristic of the spectrum that was not expected is the remarkable similarity, in this wavelength range, to the spectra of Comet Seargent and Comet West (obtained from rockets). The working hypothesis is that the ultraviolet spectra of comets may be similar, unlike the visual spectra. The visual spectra show considerable variation with heliocentric distance, and at a given heliocentric distance show variation in the strengths of the continuous emission relative to the molecular and atomic emission (as determined by the relative proportions of ices and dust in the nucleus).

Finally, we must mention a very recent result obtained by *IUE* on Comet IRAS–Araki–Alcock. This comet passed very close to the earth on May 11, 1983, and the close passage allowed the apparent detection of diatomic sulphur, S_2, for the first time in comets. This discovery may have important implications for our understanding of the physics and chemistry of the cometary nucleus.

Figure 6–16 (opposite). International Ultraviolet Explorer (IUE) *observations of Comet Bradfield on January 10, 1980. (Top) Short wavelength spectrum. The hydrogen Lyman-α line at 1216 Å is overexposed in this 2-hour exposure. Other detected species are marked. (Bottom) Long wavelength spectrum. The 3 hydroxyl (OH) bands marked are overexposed in this 2-hour exposure. Other detected species are marked. The line labeled "H" is spurious, as it is produced by the strong Lyman-α line at 1216 Å in the short wavelength spectrum above. (After figures by P.D. Feldman* et al., Nature *286: 132, 1980.)*

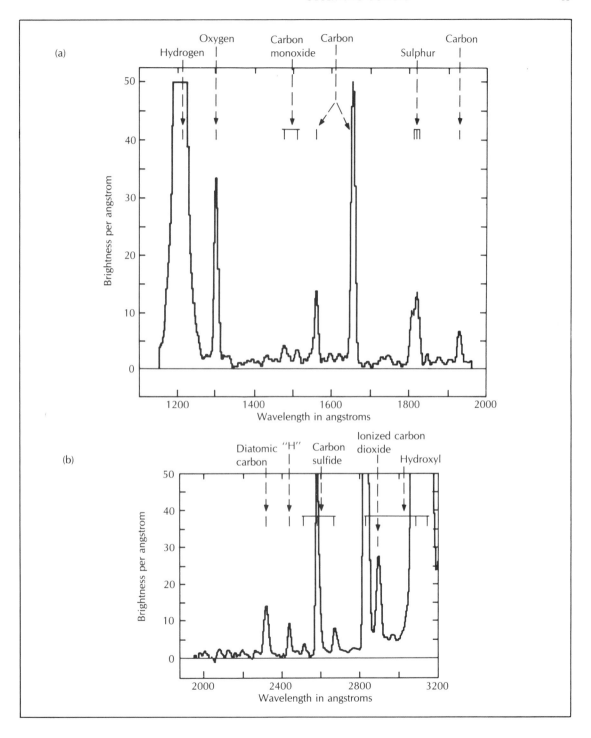

CHAPTER 7

COMETS AS SOLAR SYSTEM BODIES

DEATH OF A COMET

Periodic Comet Biela was first discovered in 1772. With its short period of roughly 6.6 years, it was observed to return several times in the early nineteenth century. Shortly after its rediscovery on its 1846 return, the comet split into two fully-developed comets. The two comets then moved off in the same orbit. As they traveled, they underwent major changes in brightness, with first one and then the other being more brilliant. In 1852 the two comets returned once again, though they were then over 2 million kilometers apart. The 1866 return was to be a very favorable one, with the comets well placed for observations; however they were never seen again.

Other comets have vanished on a single pass through the solar system. Periodic Comet Westphal (1852 IV, 1913 VI) was predicted to be very bright during its 1913 apparition. However, as it moved in toward the sun it grew very rapidly, became increasingly faint, and vanished. A comet has even been seen to plunge into the sun (see Chapter 8).

These cases should not be too surprising. After all, the gaseous parts of a comet arise because of the sublimation of the icy nucleus. Eventually, the nucleus is going to be used up. Astronomers believe that a bright comet like Halley loses a few meters of its nucleus on every pass. If the diameter of a nucleus is as small as a kilometer, then it should survive for less than a thousand passes or less than 76,000 years. If the solar system is indeed over four billion years old, then all the periodic comets should have vanished long ago, unless some process is adding new ones to the inner solar system. What might that process be?

In Chapter 5 we discussed the Oort cloud idea: that there is a large cloud of comets between 50,000 and 150,000 A.U. from the sun. Occasionally a passing star will go through the cloud and perturb

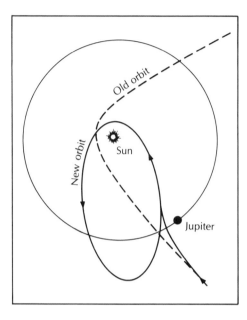

Figure 7–1. Idealized example of how a comet's single encounter with Jupiter might cause it to be captured into a short-period orbit.

a large number of comets, sending some in toward the sun. These comets will wind up with nearly parabolic orbits and periods in the millions of years. Another mechanism must be proposed for the production of short-period comets.

Short-period comets are produced from long-period comets that interact with Jupiter and the other giant planets as they pass through the inner solar system. Originally, it was thought that a single close encounter with Jupiter would suffice (Figure 7–1). In fact, single encounters can capture comets into short-period orbits. However, such encounters are very rare and cannot account for the observed number of short-period comets. Instead, it now seems that the capture of long-period comets into the inner solar system results from the accumulated perturbations of hundreds of not-so-close interactions with Jupiter and, to some extent, with the other giant planets.

It seems reasonable to suppose that some sort of steady state exists today. That is, short-period comets slowly sublime away and disappear from the inner solar system. At more or less the same rate, long-period comets are captured into the inner solar system by the effects of Jupiter. Since the number of new comets offsets the number lost, there is no net change in the number of short-period comets.

COMETS IN THE SOLAR SYSTEM

What happens to the large-particle dust debris left behind when it is released from the nucleus of a comet? All evidence points to the fact that the dust particles remain in the same orbit as the comet and eventually outline the entire path. This fact was dramatically illustrated by Comet Arend–Roland (1957 III). For several days during late April, 1957, the comet showed an *anti-tail,* a spike pointing toward the sun (Figure 7–2). Astronomers came up with a number of suggestions to explain the phenomenon: perhaps the spike was a type of material that was drawn toward the sun rather than repelled. Then it was noticed that the spike was best seen as the earth crossed the plane of the comet's orbit. What we were seeing was the orbital plane of the comet, outlined in large dust grains and illuminated by sunlight. The dust was noticeable only when seen edge-on or nearly edge-on, so that our line of sight passed through a maximum thickness of dust.

Sometimes the earth crosses not just the orbital plane of a comet

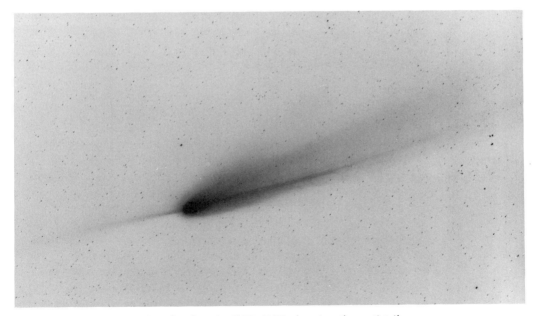

Figure 7–2. Comet Arend–Roland on April 27, 1957, showing the anti-tail, an apparently sunward spike (Palomar Observatory photograph).

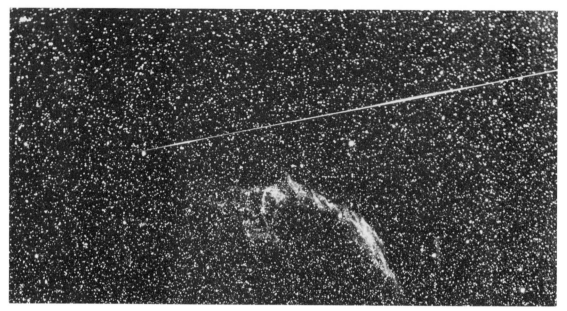

Figure 7–3. A meteor seen in the constellation Cygnus. Most meteors are believed to be caused by pieces of solid material from comets burning up as they enter the earth's atmosphere (Yerkes Observatory photograph).

but the orbit itself. In 1899 the earth crossed the orbit of periodic Comet Tempel–Tuttle (1866 I), which has a period of 33 years. A magnificent meteor shower occurred, with 100,000 meteors per hour appearing to radiate from the constellation Leo. We now call the shower the "Leonids" for that constellation. The Leonid shower is not equally spectacular every year. The best displays occur every 33 years. Therefore, the debris must be clumped up in the comet's orbit.

The Andromedid meteor shower, which occurs in November every year and appears to radiate from the constellation Andromeda, is associated with Comet Biela, which disappeared last century. If Biela had not been observed before it disappeared, then the Andomedid shower would be a meteor stream without a known associated comet. There are, in fact, many such meteor streams. Astronomers assume they all arise from comets that disintegrated long ago. The modern view is that most meteors (Figure 7–3), whether isolated (called sporadic meteors) or in meteor streams, are caused by solid debris from comets entering the earth's atmosphere. The sporadic meteors come from material originally in streams but now dispersed by gravitational perturbations of the planets.

Incidentally, it is possible to obtain spectra of the brief light flashes produced by meteors, and thus learn something about the composition of the meteoroid—the solid body in space—that produces it. Lines of metals such as sodium, magnesium, calcium, and iron have been observed, as well as some more volatile materials such as hydrogen, carbon, oxygen, nitrogen, cyanogen, and C_2. Observations of nitrogen and oxygen are tricky, because these substances make up the bulk of the earth's atmosphere, which is also heated by the passage of the material. The abundances observed are consistent with abundances found in the sun.

THE SOLAR WIND

Comets are not only intrinsically fascinating, but are also interesting as probes of the solar wind and interplanetary field. Their detailed study helps us learn about plasma processes in space.

Early in the present century, astronomers concluded that solar radiation pressure provided the repulsive force that caused comet tails to point away from the sun. However, by 1950 it was clear that there was a problem with this conclusion. The plasma tails were far too straight and oriented too nearly radially away from the sun. The ionized molecules were experiencing forces in excess of that of radiation pressure. In 1951, the German astronomer L. Biermann (Figure 7–4) hypothesized a flow of ionized material from the sun. This "wind" would interact with the ionized molecules in the comet, leading to the observed forces. The plasma tail would be like a wind sock in the solar wind. Later in the same decade, E. Parker of the University of Chicago showed that the solar corona is not a static structure, but is continuously expanding. He showed mathematically that a high-speed solar wind should exist.

Figure 7–4. Ludwig Biermann.

Interplanetary spacecraft flown after 1959 by the Soviet Union and after 1961 by the United States directly measured the solar wind. The best early observations were made by the *Mariner 2* spacecraft on its way to the planet Venus in 1962. During the three-month flight, the solar wind was always present. The average density of protons (and of electrons, since the wind is electrically neutral) was measured to be 5 protons per cubic centimeter. The wind speed varied from 319 to 771 km/sec, with an average close to 500 km/sec. The measured wind was supersonic, and in close agreement with Parker's predictions.

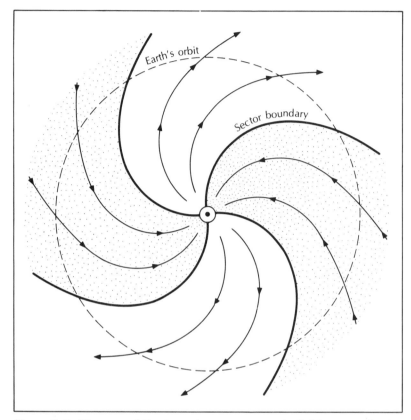

Figure 7–5. Diagram of the interplanetary magnetic field as seen in the plane of the earth's orbit, showing both the spiral shape and the sector structure, an ordering of the magnetic field into alternate "sectors" of inward- and outward-directed magnetic field.

The solar wind has an interesting effect on the magnetic field of the sun. A plasma cannot flow across lines of force in a magnetic field, but drags the field along. The solar wind drags the solar magnetic field into interplanetary space. However, because of the sun's rotation, the field lines should not be radial. In fact, the magnetic field has a spiral shape (Figure 7–5). Since the solar wind cannot cross the field lines, the plasma ions move in a path similar to the way the needle moves across a phonograph record. The plasma flows radially at just the right speed to stay "in the groove" as the magnetic structure rotates. The Mariner 2 spacecraft, and two years later the Imp 1 spacecraft, carried magnetometers to measure the interplanetary magnetic field, and verified the important sector structure of the field (Figure 7–5).

Over the intervening years, scientists have built up a more detailed picture of the solar wind—a picture for which we will not give details here. One feature is persistent high-speed streams where the wind speed can be 800 km/sec. The high-speed streams recur at 27-day intervals—the rotation period of the sun—indicating that they come from some fixed region on the sun.

All the spacecraft that have measured the solar wind have flown relatively near the plane of the ecliptic. Comets have permitted us to sample the three-dimensional structure of the wind. The very best models of the solar wind derived from comets have a radial wind speed of 400 km/sec, with small departures from pure radial flow.

Hannes Alfvén (Figure 7–6), working in Sweden and at the University of California at San Diego, put forward a theory of comet tails in 1957, in which he proposed that the interplanetary magnetic field is attached to the head of a comet as the comet moves through the field. In this theory, comet tail phenomena are dominated by magnetic processes. Support for this idea is provided by Figure 7–7, which shows a series of images of Comet Kobayashi–Berger–Milon with rays in the tail rapidly closing onto the axis of the tail. We believe the rays are magnetic field lines made visible by fluorescing cometary material being captured by the comet. Figure 7–8 is a photograph of a wire model constructed by scientists from the Soviet Union to illustrate the attachment of magnetic fields to the head of a comet. For his contributions to our understanding of electromagnetic processes in the universe, Alfvén was awarded the 1970 Nobel Prize in physics.

Figure 7–6. Hannes Alfvén.

DISCONNECTION EVENTS

A spectacular series of photographs of Halley's comet was obtained on June 6–7, 1910 (Figure 7–9). Astonomers made photographs at three observatories spread around the world: Williams Bay, Wisconsin; Honolulu, Hawaii; and Beirut, Lebanon. Since the observatories were widely separated in longitude, it was always night at one of the sites, and the comet could be followed for extended periods of time. On the nights of June 6 and 7, the tail of the comet apparently broke from the head and floated away, while a new tail formed. An even more spectacular event had been observed in Comet Morehouse in 1908, as shown in Figure 7–10.

We now believe that tail disconnections occur when a comet crosses a boundary in the interplanetary field between an inward-

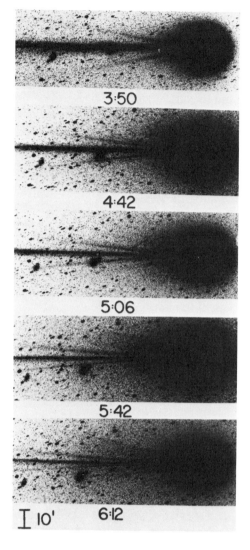

3:50

4:42

5:06

5:42

I 10'   6:12

Figure 7–7. Comet Kobayashi–Berger–Milon on July 31, 1975. This sequence of photographs shows a pair of tail rays closing onto the tail axis (see Figure 7–8). (JOCR photographs.)

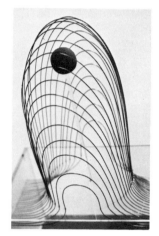

Figure 7–8. Wire model of the interplanetary magnetic field draped around the head of a comet. (Courtesy of I. Podgorny; from The Moon and the Planets *23: 323, 1980. Used by permission.)*

directed sector and an outward-directed sector (recall Figure 7–5). The rapid change in the direction of the interplanetary field cuts the cometary field somewhere in the head, and the interplanetary field sweeps the old tail away while a new tail forms (Figure 7–11).

Malcolm Niedner and John Brandt have studied a number of these disconnection events observed in the last century, and have found that most of them have been associated with the comet crossing a field reversal boundary in the interplanetary field.

Figure 7–9. A disconnection event observed in Halley's comet in 1910. The times and locations of the observations are (top) Williams Bay, Wisconsin, June 6, 15.8 hours; (middle) Hawaii, June 6, 18.5 hours; (bottom) Beirut, June 7, 7.0 hours. Notice the upper part of the tail floating away from the head (Yerkes Observatory photographs).

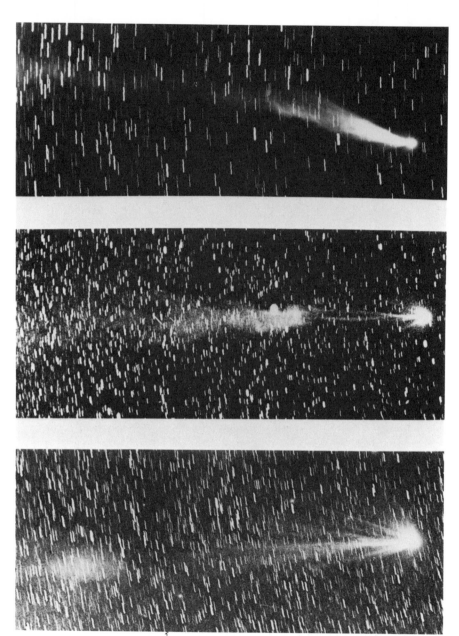

Figure 7–10. Comet Morehouse on September 30, October 1, and October 2, 1908 (top to bottom, respectively) showing the rejection and drifting away of the old tail and the partial formation of a new one (Yerkes Observatory photographs).

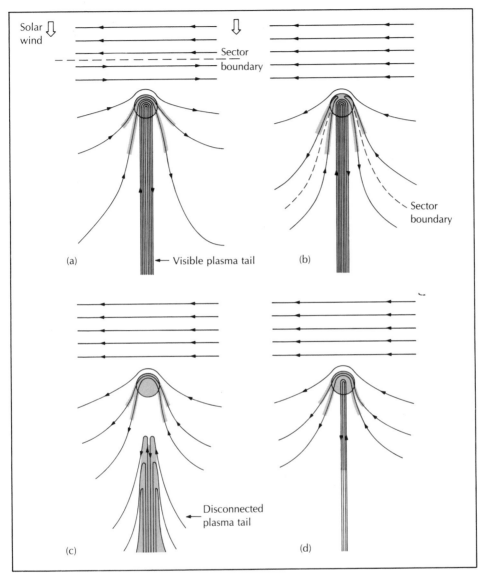

Figure 7–11. This diagram illustrates a possible explanation of disconnection events, suggested by M.B. Niedner and J.C. Brandt. When a comet crosses a sector boundary, a magnetic field of polarity opposite to the field already captured is pressed into the comet. This can result in severing the field lines draped around the comet and the tail is disconnected. The comet immediately begins to rebuild the tail, as is clearly shown in Figure 7–10 (Laboratory for Astronomy and Solar Physics, NASA/Goddard Space Flight Center, Greenbelt, Maryland).

CHAPTER 8

UNUSUAL COMETS; UNUSUAL EVENTS

On December 13, 1925, Ensor, observing from South Africa, discovered a faint comet in the constellation Reticulum. The comet grew in brightness and size throughout the remainder of December and January, reaching almost naked eye brightness. The comet passed perihelion early in February, 1926, and began to behave in an erratic fashion. Its motion departed significantly from the predicted path; what appeared to be a tail was observed 60° to 80° from the usual direction away from the sun; and, though predictions indicated that the comet should become very bright, it grew faint and diffuse and disappeared, never to be seen again. Clearly, some cataclysmic process was disrupting the usual events in this comet. Thus, Comet Ensor (1926 III), like Comet Perrine (1897 III), became one of the vanishing comets.

Over the years, comets have been observed to disappear, split, "explode," plunge into the sun, or behave in other extravagant ways. In this chapter, we will look at some of the unusual things that comets do, and in the process will learn more about their physics.

ENCKE'S COMET

Encke's comet (Figure 8–1) is remarkable for the fact that it has the shortest known orbital period: 3.3 years. The comet moves in a moderately elliptical orbit (Figure 1–4) that takes it somewhat inside Mercury's orbit at perihelion and half-way between Mars and Jupiter at aphelion. Its orbit is inclined by 12° to the ecliptic. Because of its special orbit, early attempts to calculate orbital elements met with great difficulty.

As it turns out, this comet had been discovered a number of times in the late eighteenth and early nineteenth centuries. In November, 1818, Pons discovered a comet that he was able to observe for

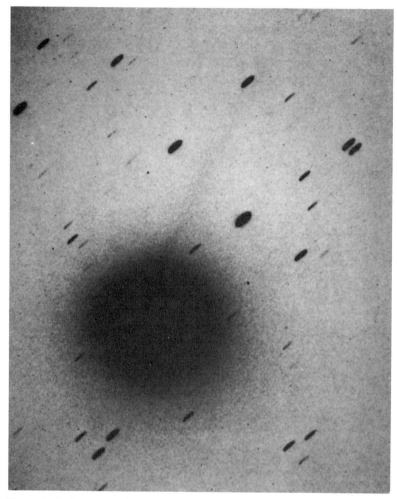

Figure 8–1. Comet Encke on January 6, 1961. Notice the weak plasma tail. (Photographed by E. Roemer, University of Arizona, Tucson, Arizona; official U.S. Navy photograph.)

almost two months. In Germany, Johann Franz Encke (1791–1865; Figure 8–2) attempted to fit a parabolic orbit to Pons' observations, as Halley had to his comet, but the attempt failed. Encke could not find any parabola that would fit the data. Fortunately, considerable progress had been made in celestial mechanics by 1818. The first minor planets had been discovered, and Gauss had devised a method for calculating their orbital parameters. Encke applied Gauss' method to Pons' observations and found the very unusual orbital characteristics

Figure 8–2. Johann Franz Encke, 1791–1865 (Yerkes Observatory photograph).

mentioned above. With a period of only a little over three years, the comet must have been seen before. Sure enough, in searching for earlier appearances, Encke found the comet had been observed by Mechain in 1786, by Caroline Herschel in 1795, and by Pons and others in 1805. In fact, Encke himself had attempted to fit a parabolic orbit to an observation prior to 1818, with limited success.

A comet that spends its time in the inner solar system, and that occasionally comes very close to Jupiter, must be subject to significant planetary perturbations. Working without the aid of a computer, Encke carried out the difficult task of calculating these planetary perturbations, and predicted his comet would return to perihelion on May 24, 1822. The comet was sighted as predicted, and has been seen at every return since—with one exception. It was not seen in 1944, in the darkest hours of World War II. In honor of his monumental work, the comet was named for Encke. It remains one of the few comets named not for a discoverer, but for the astronomer that predicted a return.

NONGRAVITATIONAL FORCES

The story of Encke's comet does not end here. Encke continued to study its orbit for several subsequent returns. He found that even after accounting for all possible planetary perturbations, the orbital period was shortening by roughly 0.1 day each revolution. This steady acceleration of the comet is called a secular acceleration. In 1823, Encke wrote a paper in which he speculated that the comet moved through a uniform resisting medium. Another speculation was that the comet passed through meteor streams, and was impeded only when it was in a stream.

In the following years other comets (P/Pons–Winnecke*) were found to show secular accelerations like Comet Encke, lending credibility to the resisting medium idea. Even so, in 1835 the German astronomer F.W. Bessel (1784–1846) pointed out that effects other than a resisting medium could lead to a secular acceleration. For instance, Bessel noted that the jets of material observed moving outward from the nuclei of comets could create a rocket effect, to accelerate the comet. Depending on the direction of the rocket, it could lead to either an acceleration or a deceleration. In the 1930s two comets (P/Wolf and P/d'Arrest) were observed to exhibit secular decelerations, ruling out the resisting medium theory and lending credibility to Bessel's idea.

The modern idea of nongravitational forces that cause secular changes in the periods of comets follows Bessel's proposal. Heat from the sun causes the nucleus to sublimate on the side nearest the sun. If this were the only thing that happened, there would be no secular period change. A rocket effect radially away from the sun does not cause secular changes. However, if the nucleus also rotates, the point that has been heated by sunlight will always move a small distance while the resulting sublimation takes place, and the rocket effect will not be exactly radial. Depending on the direction of rotation of the nucleus, the rocket effect can cause either an acceleration or a deceleration; see Figure 8–3.

SUN-GRAZING COMETS

Comet 1843 I was a brilliant comet. On February 28 it was observed in broad daylight 4° from the sun. At least ½° of its tail could be seen against the bright daytime sky. According to estimates by various observers, the comet was up to 40 times brighter than the planet Venus. What was most spectacular about this comet, however, was its dis-

*Periodic comets are frequently noted by the symbol P/ before their name.

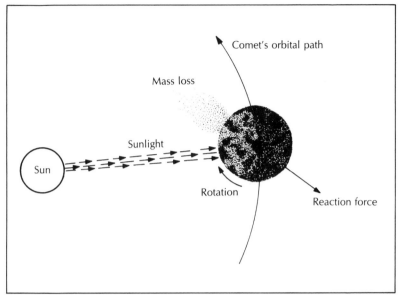

Figure 8–3. Diagram showing the origin of the nongravitational force or "rocket effect."

tance from the sun at perihelion passage. It was 0.0055 A.U. or 825,000 km from the center of the sun. Since the sun is 700,000 km in radius, the comet was a mere 125,000 km or 18% of the solar radius above the surface—a veritable near miss. Comet 1843 I is a member of a group of comets known as the sun grazers.

When Comet 1843 I was at its closest point to the sun, it was well inside the corona. (Of course, depending on the definition of the corona, even the earth is inside it.) The density of the corona is about 10^8 particles per cubic centimeter, and the temperature about 2,000,000°C. At such low densities, the coronal plasma does not contain enough energy to melt the comet's nucleus. However, the comet received over 30,000 times more radiant energy in the form of sunlight than it would at the earth's distance from the sun. An ideal body known as a black body* sitting 125,000 km above the sun's surface would quickly heat up to 5000°C—a temperature that would vaporize any known substance. The comet survived for two reasons: it moved very rapidly, and the process of sublimation carried the heat away quickly.

*A black body is a body that absorbs and then re-emits all electromagnetic radiation that hits its surface.

Figure 8–4. Woodcut of the Great Comet of 1843 as seen from Paris on the night of March 19. (From The World of Comets, *by A. Guillemin, London, 1877.)*

It was also probably quite a large comet. As it moved away from the sun it developed a very long tail (Figure 8–4). By some estimates, the tail achieved 2 A.U. in length, the longest known.

Until recently the comet of 1843 was the closest sun grazer known. Then a comet was seen that appears to have plunged into the sun.

THE COMET THAT HIT THE SUN

Comet Howard–Koomen–Michels (1979 XI) hit the sun on August 30, 1979. Knowledge of this comet and of its fate was acquired in a most unusual manner. The observations were made by an instrument

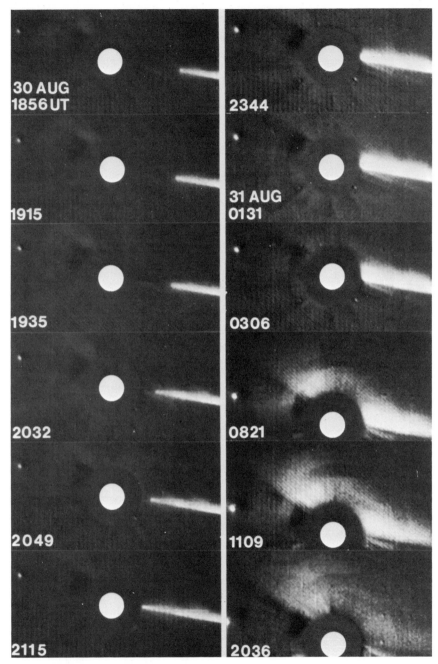

Figure 8–5. Sequence of photographs showing comet Howard–Koomen–Michels hitting the sun on August 30 and 31, 1979. (Photographs courtesy of Naval Research Laboratory, Washington, D.C.)

designed by the U.S. Naval Research Laboratory to monitor the solar corona and flown on a U.S. Air Force Space Test Program satellite.

The instrument, a coronagraph, records the corona from a projected distance of 2.5 solar radii out to 10 solar radii. The inner area to a distance of 2.5 solar radii is blocked by an occulting disk used to create an artificial "eclipse" of the sun's disk. Thus, the blinding photospheric light is prevented from entering the coronagraph. The shadow of the occulting disk is visible in the observations shown in Figure 8–5.

The sequence of observations clearly shows the comet approaching the sun, passing behind the occulting disk, and apparently disintegrating. A major portion of the corona becomes bright, presumably from cometary material blown from the area of the comet's disintegration by solar radiation pressure. The large extent of this event can be emphasized by recalling that the occulting disk seen in Figure 8–5 has a projected radius of 2.5 solar radii. The enhanced brightness extended to between 5 and 10 solar radii, or millions of kilometers. The comet would have been seen in the observations if it had reappeared from this "close encounter." This comet was the first discovered from a spacecraft and the first observed to collide with the sun.

Although the arc through which the comet was observed is very short, it was possible to derive an orbit of modest accuracy. The part near the sun is shown in Figure 8–6 for two cases, corresponding to perihelion distances of 1.1 solar radii (dashed line), that is, a "near-miss," and of 0.75 solar radii (dotted line), which produces impact with the photosphere at the location marked with an asterisk at 22^h29^m Greenwich mean time. The orbital elements reveal another curiosity. Comet Howard–Koomen–Michels may be a member of the Kreutz group of sun grazers. This group of comets is thought to have originated from a proto-comet that may have split during an earlier close encounter with the sun. Well-known members of this group are the Great Comet of 1882 (this chapter) and Comet Ikeya–Seki (1965 VIII), Plate 7. Several other sun hitters have been discovered by the NRL group.

SPLIT COMETS

Comet 1826 I was discovered in February, 1826, by the German astronomer W. von Biela (1782–1856), and was observed for almost three months. The orbit calculated from these observations showed that the comet had been observed twice before, as Comet 1772 and as Comet 1806 I. In the latter appearance it had reached naked eye

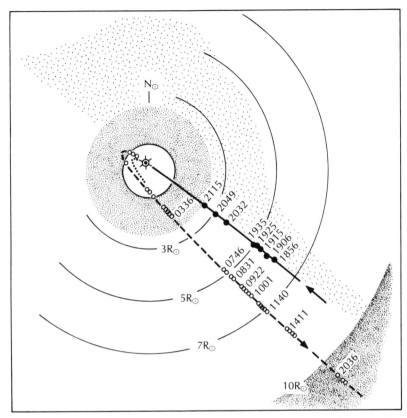

Figure 8–6. Orbit of Comet Howard–Koomen–Michels near the sun. The north pole of the sun is marked by the symbol $N_\odot$ and the projected distances are measured in solar radii ($R_\odot$). (Courtesy of Naval Research Laboratory, Washington, D.C.)

brightness for a short time. The comet, now called Comet Biela, has a relatively short period of 6.6 years. Based on the 1826 orbit, the comet was predicted to return in 1832, and it showed up on schedule as Comet 1832 III. The next two returns were unfavorable, and the comet was not observed.

In 1845 the comet was once again observed—the fifth recorded appearance. However this appearance proved to be more fateful and more fascinating than previous ones. On the night of January 13, 1846, a faint satellite comet was observed a small distance from the main comet. Both comets had tails, which were parallel to one another. Over the next month the fainter of the two comets increased in brightness and finally became brighter than the "main" comet. The situation then reversed and the main comet became the brighter one again. In addition, the main comet grew a second tail and a luminous bridge of material joined the two comets. Throughout the time that the two components changed brightnesses back and forth, they moved in parallel paths about 250,000 km apart. As the two comets moved away from the sun and the earth, the satellite comet faded first, then the main comet.

Everyone was waiting for the 1852 return. As described in Chapter 7, both comets returned at the predicted time, though they were over 2 million kilometers apart. Once again the two comets took turns as the brighter of the pair. On at least one occasion a bright jet was seen between the two heads. Otherwise, the apparition was not unusual. The 1866 return was to be a very favorable one; the comet was well situated relative to the sun and the earth. However, it did not return and it has not been seen since. Or has it?

In November, 1872, a particularly spectacular meteor shower was observed radiating from a point—the *radiant*—on the sky in the constellation Andromeda. The shower, known as the Andromedids for the constellation, is an annual event, known to be associated with Comet Biela. When the shower was particularly intense, E.F.W. Klinkerfues (Chapter 2) concluded that the earth must have encountered the main comet body, and decided the comet could be seen moving away from us at a point on the sky (near the star theta Centauri) opposite the radiant point in Andromeda. Klinkerfues sent a telegram to his friend N.R. Pogson in Madras, India (which was far enough south for easy viewing of Centaurus) urging a search. Pogson found a comet near the proper location, and observed it for several days. Over the years there has been much debate whether Pogson saw Biela's comet or another comet that just happened to be nearby (recall the strange case of Comets Perrine 1897 III and 1898 I, told in Chapter 2). Astronomers believe that Biela should have been seen three months earlier. If Pogson did see Biela, it was truly the last sighting; if not, it stands as another incredible coincidence.

A more interesting question than Pogson's observation is, What happened to Comet Biela in the first place? The answer seems to be simple: the nucleus split up. About a dozen cases have now been observed, one of the most recent being Comet West (1976 VI). In March, 1976, West was observed to have four nuclei (Figure 8–7). The nuclei remained close together in one coma, so that multiple comets did not result.

In most cases there is no apparent reason for a comet to split up, and we can only assume that the process of sublimation might lead to a highly irregular body that simply does not have the strength to hold together. One exceptional case is the great Comet 1882 II. The September comet of 1882 was another of the sun-grazing comets. On

Figure 8–7. Nuclear splitting in Comet West. The dates shown are (left to right) March 8, 12, 14, 18, and 24, 1976 (New Mexico State University Observatory).

September 17, 1882, the comet passed perihelion and skimmed closer than 500,000 km to the solar surface. After perihelion the nucleus was observed to be multiple: five nuclei appeared to be on a nearly straight line. It is possible in this case that the tremendous tidal forces experienced by the comet as it skimmed by the sun tore the nucleus apart. Again (recall the Great Comet of 1843) the comet developed into a spectacular object.

Zdenek Sekanina (Figure 8–8), working in the United States, has studied the relative motions of the fragments of several of the well-observed split comets, including Comet West. He concludes that the relative motions may result from different nongravitational forces acting on the various fragments.

Figure 8–8. Zdenek Sekanina.

GREAT COMETS

There are a handful of comets that deserve the title "great comets." The sun-grazer Comet 1882 II, described earlier, is one of the most spectacular of these (Figure 8–9). The comet was first observed, after it reached naked eye brightness, by sailors aboard an Italian ship in the southern hemisphere. Within two weeks after it was first observed, the comet was so bright that it could be seen in broad daylight. One report stated that on September 7 the nucleus of the comet was only a little fainter than the full moon. Comet 1882 II remained a naked eye object in southern skies into March 1883.

It is interesting to note that there appear to have been many more really spectacular comets in the nineteenth century than in the twentieth, and the most spectacular great comets were also sun grazers: Comets 1843 I, 1880 I, 1882 II, 1887 I, and 1965 VIII (Ikeya–Seki,

Figure 8–9. The Great Comet of 1882 as photographed by David Gill in South Africa on November 13. This 80-minute exposure was one of the first photographs that successfully recorded a comet's over-all appearance and helped demonstrate the potential of photography in cometary studies (Royal Astronomical Society, London).

Plate 7). There can be little doubt that the brilliance of these comets was due in part to their nearness to the sun.

Comet 1880 I was interesting because it was first seen in the evening of January 31, 1880, as a naked eye object. Only a 40° long tail was seen above the horizon. The comet had just passed perihelion, and its head was still quite near the sun.

These few examples show the extreme variety of phenomena that comets can exhibit. In Chapter 9 we will look at a number of the fascinating stories that surround comets.

STORIES AND LEGENDS

1066

Modern astronomers have looked carefully at reports of comets in ancient Chinese records, in Greek and Roman writings, and in medieval European records. Again and again they find mention of a bright comet that, because of the dates it was recorded and its motion on the sky, must have been Comet Halley at an earlier return. The earliest such recorded return was in 240 B.C. No known records exist for the apparition in 164 B.C. However, mention can be found of a comet that must have been Halley for every return from 88 B.C. to the present—an unbroken string of 27 returns covering more than two thousand years.

When we consider the large number of returns of Halley that have been observed and the long period of time involved, it is not too surprising to find that the comet has been visible in the sky during many historically important events. Some individuals, either innocently or maliciously, were quick to blame these events on the comet's appearance, adding fuel to the fire of the belief that comets are portents of disasters.

One of Halley's returns occurred in the year 1066, when the comet was visible throughout the spring season. That was a fateful year in European history. Harold was crowned King of England in January, after the death of Edward. William, Duke of Normandy, also had some interest in the throne. Apparently, Edward had promised to bequeath the throne to William in return for William's friendship, and then conveniently forgot the promise. A number of other intrigues involving Harold's exiled brother Tostig and King Harald Hardrada of Norway all led to the famous battle of Hastings in October, 1066, in which Harold was hacked to pieces by the Norman invaders, leaving William as undisputed King of England.

The events of the Norman Conquest are chronicled in the so-called Bayeux tapestry. The tapestry, which is actually an embroidered linen strip, was supposedly executed by William's queen, Matilda, and her ladies in waiting. In one of the sixty scenes depicting events from the battle preparations to the death of Harold, a comet—Halley's comet—is shown in the sky (Plate 3). Depending on your point of view, you can regard the apparition as a bad omen for Harold—as the tapestry does—or as a good omen for William the Conqueror and his Norman troops.

CATASTROPHES AND COMETS

In 1696, William Whiston (1667–1752) published a treatise entitled *A New Theory of the Earth, from its Origin to the Consummation of All Things, Wherein the Creation of the World in Six Days, the Universal Deluge, and the General Conflagration, As Laid Down in the Holy Scriptures, Are Shewn to be Perfectly Agreeable to Reason and Philosophy.* Whiston's thesis in this treatise is that catastrophes such as the Deluge set forth in the Bible occurred as a result of an encounter between the earth and a comet.

After predicting the 1759 return of the comet of 1682 (Halley's comet), Halley began to look for other periodic comets. He decided— incorrectly as it turns out—that the comet of 1680 orbited the sun with a period of 575 years. Whiston noted that that comet would have appeared in 2920 B.C., the date given by the biblical chronologies of such men as Archbishop Usher as the time of the Deluge. Whiston argues that the flood was caused when the tail of the comet impacted the earth. (In Chapter 10 we will see that an encounter with the tail of Halley's comet in 1910 produced no noticeable results.)

In his book *Worlds in Collision,* published in 1950, Immanuel Velikovsky argued in a vein similar to Whiston's. Velikovsky attempted to explain such biblical events as the parting of the Red Sea at the time of the Israelite exodus from Egypt as due to a comet that erupted from a volcano on Jupiter, encountered the earth, and finally became the planet Venus.

1759

The most eventful apparition of Halley's comet was probably the 1759 return, the first to be predicted. We discussed its influence on science

Figure 9–1. John Win-throp, 1714–1779 as painted by J.S. Copley in 1773. (Courtesy of the Harvard University Portrait Collection, Cambridge, Massachusetts.)

earlier (Chapter 4), but entirely in terms of European science. The return also generated a significant amount of interest among colonists in America. American astronomers did not make any major contributions to the birth of celestial mechanics, or to astronomy, in the eighteenth century. Yet there were several colonial colleges where up-to-date astronomy was taught. Most notable was Harvard, where John Winthrop (Figure 9–1) taught and studied.

Winthrop was born in Boston in 1714. He graduated from Harvard in 1732, where he showed himself to be a brilliant student. In 1738 at the age of 24, he was appointed to the Hollis Professorship of Mathematics and Natural Philosophy, a post he held for the next forty years. Winthrop's scientific career was highly successful. It is said that he, more than anyone else, established colonial America as an independent scientific arena. He began to correspond with English astronomers, and reported a number of important observations to the Royal Society. He was fully aware of the great advancements being made by Newton, Halley, and other European scientists.

When the 1759 return of Halley's comet occurred as predicted, Winthrop presented a series of lectures to his students at Harvard. The lectures, which have been published, presented as scientifically accurate a picture of comets as could be expected for that time. After his scientific discussion, however, Winthrop turned his attention to moral theology, and concluded that the motion of comets demonstrates the orderly design of the Creator. He also followed the pattern set by William Whiston and talked of the cometary theory of the Deluge. Nevertheless, Winthrop also wrote: "To be thrown into a panic whenever a comet appears, on account of the ill effects which some few of them might possibly produce, if they were not under proper direction, betrays a weakness unbecoming a reasonable being."

1812

In December, 1811, a series of earthquakes began that rocked over three hundred thousand square miles of eastern North America. The great New Madrid Earthquake was just one of the many events of late 1811 and 1812 that followed quickly on the heels of the Great Comet of 1811. First observed from America in late 1811, the comet was described as bright and slightly smaller than the full moon, including its tail. Newspapers in the young republic picked up on the comet and predicted that it was an omen of evil times. Sure enough, a string of disasters, natural and otherwise, followed.

Though the epicenter of the earthquakes was near a small town on the Mississippi River—New Madrid, Missouri—the numerous shocks were felt as far away as New York and Florida. According to one source, Richmond, Virginia and Boston were shaken so violently by the December 16 shocks that church bells rang. It is said that in some places the current in the Mississippi River flowed backward.

The great naturalist John James Audubon was in Kentucky when the first tremor struck. Initially he thought the roar was a tornado and headed for shelter. Audubon and others reported strange darkenings and brightenings in the sky.

Jared Brookes, a native of Louisville, Kentucky kept accurate records of all the shocks that he experienced. Between December, 1811, and May, 1812, he tabulated over two thousand separate tremors. Based on the historical evidence, Charles Richter (inventor of the Richter scale used to measure the strength of earthquakes) has estimated that there were at least three severe shocks that exceeded magnitude eight on his scale. The New Madrid Earthquakes thus stand as one of the most severe, if not the most severe, series of quakes recorded in U.S history.

The New Madrid Earthquakes were not the only disaster to take place in 1811 and 1812. On December 26, 1811, a fire broke out in the new theater in Richmond, which was packed with people. Governor George Smith and almost eighty others perished. The incidents leading to the War of 1812 were moving inexorably forward: the Battle of Tippecanoe; the *Guerrière* incident in which the British impressed an American seaman from an American vessel onto their warship *Guerrière*. To round out the problems, severe weather plagued the young republic. All told, 1812 was not a good year.

An interesting contrast to the disasters of 1811–1812 comes from the world of wine. The year 1811 produced a particularly good vintage. In honor of the great comet, the wine was referred to as *vin de la comète* (comet wine). Not everything from the year was bad!

TUNGUSKA

An amazing event occurred in central Siberia in the morning hours of June 30, 1908.* A huge explosion took place in the basin of the Podkamennaya Tunguska River. The explosion dwarfed the tremendous eruption of Mt. Saint Helens in May, 1980. The seismic station at Irkutsk, almost 900 kilometers from the explosion, registered an earthquake associated with the blast, and meteorological stations around the world measured an air blast wave that traveled twice around the globe. Unusual bright clouds were observed over western Asia and Europe for several nights following the explosion.

The peasants who lived in the sparsely inhabited area provided scattered reports of a bright object seen in the sky moving almost vertically downward. As the object "approached the ground it seemed to be pulverized, and in its place a huge cloud of black smoke was formed and a loud crash, not like thunder, but as if from the fall of large stones or from gun-fire was heard."† This and other reports, from an area 1,000 km in radius, convinced scientists that a giant meteorite had fallen.

Despite the obviously spectacular nature of the Tunguska Event, no attempt was made to find and visit the site until after the Revolution. In 1921, L.A. Kulik of the Mineralogical Museum of the Russian Academy of Sciences led an expedition into the area. This expedition never reached the actual fall site, which was 700 km from the nearest

---

*Before the Revolution of 1917, Russia did not use the Gregorian calendar. However, we use Gregorian or "new style" dates here.
†E. L. Krinov, *Giant Meteorites*, New York: Pergamon, 1966, p. 128.

railway line. Kulik did, however, succeed in locating a number of eye-witnesses. Their stories tell of a truly catastrophic event. One witness described how his hut was flattened to the ground and its roof blown away by the wind. His reindeer all ran away and he himself was deafened. The reports, including the seismic information, permitted the putative impact point to be located with some precision.

The account of Kulik's first expedition directly to the site in 1927 is truly an adventure story. The expedition reached Vanovara, a trading station roughly 75 kilometers from the impact point, in late March. The snow was so deep and the forest so nearly impenetrable that it took several difficult attempts to reach the site, including rafting down rivers swollen by the beginning of the spring thaw. Finally, at the end of May he reached the Tunguska fall site.

Kulik describes the center of the fall as a "vast caldron surrounded by an amphitheater of ridges and isolated summits." The forest inside and outside the caldron was completely flattened, with the fallen trees pointing away from the impact point (Figure 9–2). Within the caldron were "dozens of flat crater holes exactly like lunar craters." They ranged from ten to fifty meters across and up to four meters deep. Unfortunately, dwindling supplies forced Kulik to leave the area before he completed his investigations.

Subsequent expeditions have seriously questioned Kulik's original assumption that the caldron and holes were caused by meteorite impacts. No meteorites of any significant size have been found in or around the caldron. However, the caldron does coincide with the location of the earthquake associated with the event, and the felled trees all point radially away from it. Furthermore, extensive searches in the area have turned up no other potential impact site. All evidence collected from the many expeditions to the site leads to the conclusion that a body entered the earth's atmosphere at a shallow angle and exploded several miles above the surface. The destruction on the ground was not caused by the impact of a solid body, but by the blast wave from an airborne explosion.

In 1962 an expedition was sent into the area once again. However, this time the scientists had a helicopter at their disposal. Soil samples were collected over a large area to search for extraterrestrial dust particles. The investigators discovered a narrow tongue over 250 km long, extending northwest from the explosion site, that was enriched in meteoritic dust. All eyewitness reports indicate the body came from the east–southeast, consistent with the pattern of dust.

The big question still remains: what was the body that caused the Tunguska Event? Several implausible suggestions have been put forth over the years: a miniature black hole, an alien spacecraft, an extrater-

Figure 9–2. Aftermath of the Tunguska Event. Here trees are shown blown down approximately 8 kilometers from ground zero (Sovfoto).

restrial nuclear bomb. Good arguments can be found against each of these suggestions.

A more reasonable thought that was put forward several years ago is that the body was a small comet. There are several arguments in favor of this idea. First, the direction of motion means the body must have been moving in a comet-like orbit. The Czech astronomer Lubor Kresak has attempted to reconstruct the orbit based on the path through the atmosphere. He concludes that the body was a large chunk of Encke's comet. The unusual meteorological phenomena observed after the event—the bright skies over western Asia and Europe—could have been caused by the dust associated with the comet's tail. If the body was cometary, then it was moving at about 45 km/sec (100,000 mi/hr) when it hit the atmosphere. We know roughly how much energy was released in the explosion from a study of the destruction it caused. (We have also seen the results of a number of airborne nuclear explosions of known magnitude.) If we assume that this energy was produced entirely by a loosely compacted body decelerating rapidly, we can estimate its mass to be around 50,000 tons. Assuming a density typical for cometary nuclei (1 gm/cm^3), we can estimate the size of the body. The result is perhaps surprising at first. The object was only 40 meters in diameter. However, we must remember that a body moving at 45 km/sec has as much energy per gram in the form of kinetic energy as TNT has per gram in chemical energy. A 40-meter chunk of TNT would have a lot of energy!

The final question might be, Why wasn't the comet seen in the sky before it exploded? The answer is twofold. Its nucleus was so small that it would have been very faint, and its direction of motion at impact was almost exactly away from the rising sun, so it would have been lost in sunlight. This cometary hypothesis has been widely accepted. However, several scientists have recently completed extensive investigations that do not support a cometary origin. Zdenek Sekanina, for instance, has carefully analyzed all the evidence currently available and concludes that the body remained in one piece until it reached a point a few kilometers above the earth's surface; then it suffered an enormous terminal explosion. If the body had been a comet, he asserts, it would have been too fragile to stand up under the dynamical loads it would experience deep in the atmosphere, and it would have exploded much higher in the atmosphere. He feels the body was a small asteroid.

Should you stay awake nights worrying about being hit by another such body? Our advice is, don't. On the basis of known numbers of comets in the solar system, scientists have estimated that an impact with a comet will occur every million years or so. Sekanina estimates that there might be a small asteroid impact every 2,000 to 12,000 years. Since only a tiny fraction of the earth's surface is densely populated, an impact is likely to occur on an area with a low population, such as Siberia, or on an ocean. That is not to say that a comet could *not* hit Washington, D.C. tomorrow. The destruction would be terrible—over three million people killed, uncounted billions of dollars worth of homes and businesses destroyed. Still, the chances are so small that it is not worth worrying about.

CHAPTER 10

THE PROMISE OF HALLEY

California's Mount Whitney, at 14,494 feet, is the highest peak in the contiguous 48 states. It towers a mere 60 feet over Colorado's highest peak, Mount Elbert. It is a fairly easy climb to reach the top of Mount Whitney today. However, in 1910, when Mr. G.F. Marsh of Lone Pine, California, a town of under 2000 souls at the foot of Whitney, decided to climb the mountain to observe both the total lunar eclipse of May 23 and Halley's comet (Figure 10–1), he was faced with a difficult day-and-a-half trip. He did make it to the top, and he set up camp. Here is the rest of the story in his own words.

> I fixed up good for the night, read the weather report of 23 below zero, and fixed up a signal fire of old paper and chips so I could signal at night, but it started to cloud up and looked like storming. Heavy banks of haze formed to the west and at sundown the whole sky was cloudy and I thought it was all off. At 7:30 P.M. it cleared and I lit my fire. Clouds drifted across but I got an answer from Lone Pine.
>
> I was getting anxious about the moon and the comet when the clouds began to break. I commenced to look for the comet but the clouds bothered. Pretty soon the clouds raised and I saw the moon about half covered, and in watching so close I almost forgot the comet. I watched the moon until it was almost covered . . . then all of a sudden the comet showed in plain view. The cloud had passed by and the moon was dark. The comet was farther west than I expected. It was a good deal larger than I expected and of a milky color, but quite bright and the tail streamed out for a long distance and was very beautiful. It seemed like a great horse's tail streaming out. The comet seemed to travel very fast. In my excitement I forgot all about the time . . . . The sky up high was perfectly clear, but low down great banks of fog were rolling. I watched the comet until it dove into a fog bank to the west and was gone, but the tail shone out for quite a while.*

*Popular Astronomy 19:578 (1911).

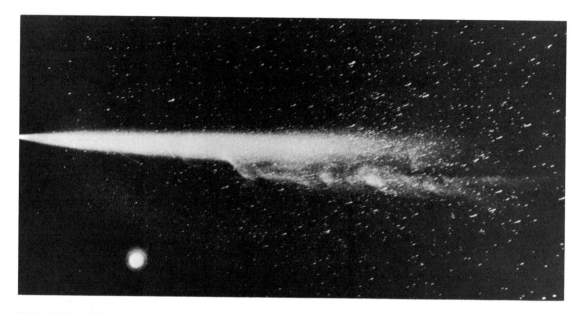

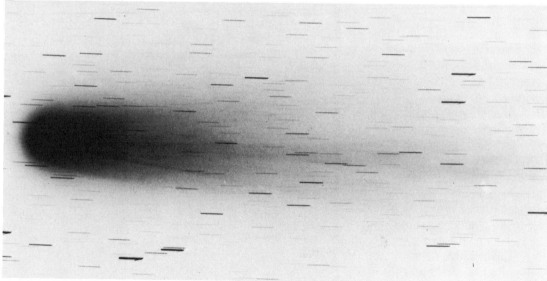

Figure 10–1. (Top) The well-known view of Halley's comet obtained on May 13, 1910. The comet stretches over 40° in the sky. The bright object in the lower left corner is the planet Venus (Lowell Observatory photograph). (Bottom) Halley's comet on May 23, 1910, the day Marsh climbed Mt. Whitney to observe it (Yerkes Observatory photograph).

We can just imagine the thrill of the sight that Mr. Marsh saw. The sky seen from the 14,000-foot mountain is crystal clear, and in 1910 it was unaffected by air pollution and city lights. In 1985–1986 when Halley returns, it will be more difficult to find any spot on earth where such a magnificent view is possible. Like Mr. Marsh, we will have to seek some locale away from large cities and with clear, dark skies to see Halley in all its glory. To appreciate what we might see, let us take a look at the last two appearances of Halley's comet.

COMET HALLEY, 1835

Between the historically important 1759 appearance of Halley's comet and its 1835 return, celestial mechanics had developed to a fine art. Also, in 1781 William Herschel discovered the giant planet Uranus,

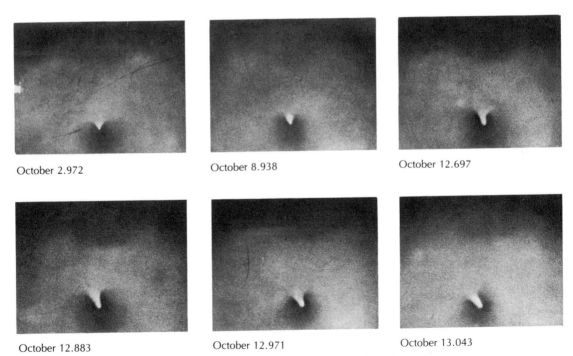

October 2.972   October 8.938   October 12.697

October 12.883   October 12.971   October 13.043

Figure 10–2. Bessel's drawings of Halley's comet in 1835 showing the region surrounding the nucleus. (From the Atlas of Cometary Forms, *by J. Rahe, B. Donn, and K. Wurm. NASA SP-198, 1969.)*

Figure 10–3. Friedrich Wilhelm Bessel, 1784– 1846 (Yerkes Observatory photograph).

which has significant perturbation effects on the comet because Halley spends much of its time near aphelion, which is near Uranus' orbit. Detailed predictions of the 1835 return were worked out by M.C.Th. de Damoiseau and Count de Pontécoulant. They, and others, predicted the comet would reach perihelion within a week or so of November 12.

The search for the comet began late in 1834, but it was not recovered until August, 1835, by astronomers at the Collegia Romano in Italy. The comet reached naked eye brightness in late September, as it passed the earth on its way in toward perihelion. The comet passed perihelion at 10:53 G.M.T. on November 16, 1835, roughly four days after the most detailed calculations had predicted. After passing near the sun, the comet once again became a naked eye object, primarily in the southern hemisphere. During this return the comet was extensively observed (Figure 10–2) by F.W. Bessel (Figure 10–3), and the study of the physical nature of comets was begun.

Reports of the visible appearance of Halley in 1835 show that the return was a favorable one. The comet was quite bright and exhibited an extensive tail. E. E. Barnard (Figure 2–7), one of the great observational astronomers of the late nineteenth and early twentieth century, has written extensively of the 1910 appearance of the comet. He points out in his papers that he tried to collect other reports of the 1835 appearance, but found the published material surprisingly meager.

COMET HALLEY, 1910

Celestial mechanicians once again carried out extensive calculations of the motion of Halley's comet in an attempt to predict its motion while it was near the earth and the sun. The calculations all led to the prediction that the comet would pass perihelion sometime on April 17, 1910. The comet reached perihelion on April 20 at 4:16 G.M.T. Once again, it was several days late. Modern calculations show that the comet has been an average of slightly more than four days late arriving at perihelion on each of the past seven revolutions. Like Comet Encke, Halley is subject to nongravitational forces that lengthen its period— that is, they cause a secular deceleration—by 4.1–4.4 days each revolution.

The 1910 appearance of Halley had many interesting aspects (Figure 10–4) and May, 1910, was particularly exciting. The earth passed through the comet's tail, the comet passed directly between the earth and the sun and transited the solar disk, and on May 6 the earth encountered fragments of the comet, causing a meteor show. Let us look at each of these events in turn.

E.E. Barnard began a search for the comet in the fall and winter of 1908–1909 at Yerkes Observatory in Williams Bay, Wisconsin. He searched both photographically, using photographic telescopes, and visually, using the famous 40-inch refractor, with no success. The comet was first seen photographically at an observatory in Heidelberg, Germany, on September 11, 1909. Barnard subsequently followed it regularly until it passed behind the sun in February, 1910. Throughout this period the comet was a faint, diffuse object visible only with the largest telescopes. Halley reappeared from behind the sun in April, and rapidly approached the earth, increasing in brightness.

Barnard reports that the comet showed a nucleus within a nucleus on May 26. The inner nucleus was stellar in appearance, but the outer one had a measurable size. The inner nucleus was probably the false or photometric nucleus observed in many comets, while the outer nucleus was probably an expanding halo.

On May 4, the tail was about 15° in extent. It stayed that length until the 9th, then it grew rapidly, reaching 53° by May 14, 107° by May 17, and 120° by May 18. As the comet whipped in front of the sun, the tail passed over the earth. The extent of the tail remained at about 50° throughout the rest of May, then began to lessen through June.

On May 24, Barnard described the comet as "quite bright to the naked eye, with traces of tail. It was bluish white and a striking object.

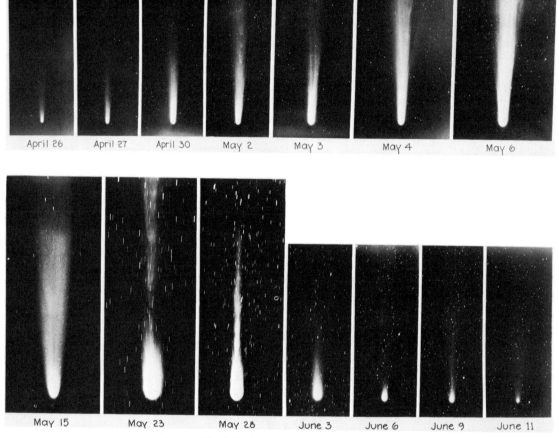

April 26 April 27 April 30 May 2 May 3 May 4 May 6

May 15 May 23 May 28 June 3 June 6 June 9 June 11

Figure 10–4. Halley's comet in 1910 (Mount Wilson and Las Campanas Observatories, Carnegie Institution of Washington photographs).

The head was large and hazy and about 15' or 20' in diameter. The nucleus resembled a first magnitude star in the haze.'' On May 29 the tail was about 52° long, but quite faint. The comet remained visible to the naked eye through the first week in June, but was overwhelmed by a waxing moon after about June 9. It was observed photographically later. On May 6, the earth passed as close as possible to Halley's orbit.

If the comet had been at the point in its orbit closest to the earth at that moment, then we would have had the closest possible approach between the two bodies. In fact, the comet was still approaching that point as the earth went by, and the closest approach between the two bodies occurred on May 18. On the morning of May 6, as the earth passed the comet's orbit, it did encounter some debris left by the comet's previous passes. Observers reported seeing meteors radiating from a point 30–40 degrees southwest of the comet, near the star eta Aquarii. Thus the meteor shower is called the eta Aquarids.

On May 18, Halley's comet passed directly between the earth and the sun. Unfortunately, the transit occurred after the sun had set throughout North America, so it was not observed here. Extensive observations were carried out at the Moscow Observatory, and were reported (in French) in the journal *Astronomische Nachrichten* for 1911. Observations began about 20 minutes before the beginning of the passage. The observing conditions were very good, and the details of the solar surface—solar granulation, faculae, and sunspots—stood out clearly. However, throughout the hour that the comet passed across the solar disk, no trace of it was seen. This observation tells us something about the size of the nucleus. Under the favorable observational conditions that prevailed in Moscow at the time of the transit, the nucleus should have been seen as a black spot if it were as large as a tenth of an arc second in diameter. During the transit, the comet was 24,000,000 km from the earth. If the nucleus were over 100 km in diameter, it would have been a 0."1 dot silhouetted against the solar disk and it should have been visible.

The coma should have covered the entire solar disk during the transit, but because the coma is so tenuous, it is not surprising it was not seen. Astronomers at the Moscow Observatory examined the solar spectrum during the transit and found no detectable change from the undisturbed spectrum. Thus, the transit was an interesting and informative event that told us something about Halley's comet by showing nothing.

The comet was observed in the morning sky the day before the transit and in the evening sky immediately after the transit. In the process of slipping between the earth and the sun, the comet's tail passed over or very near the earth (Figure 10–5).

The encounter between the earth and the tail caused much trauma, even though once again nothing untoward happened. The fear was aroused by the fact that one constituent of the tail is cyanogen, a deadly poisonous gas. An enterprising charlatan manufactured so-called comet pills, which he peddled to poorly educated people to ward off the evil effects of the comet. We have carried out a calcula-

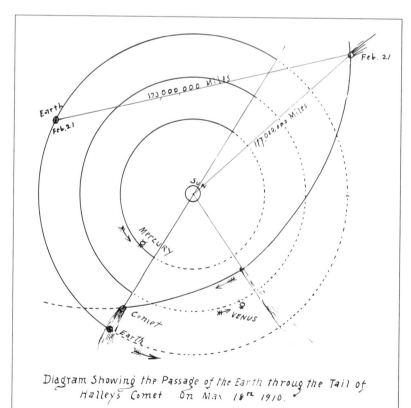

Diagram Showing the Passage of the Earth throug the Tail of
Halleys Comet On May 18ᵗʰ 1910.

Figure 10–5. Barnard's work on Halley's comet for May 18, 1910. (Top) His drawing of the tail stretching across the sky on the morning of May 18. (Bottom) His drawing showing the geometrical circumstances of the encounter (Yerkes Observatory photographs).

tion that shows how tiny the risk is from the poisonous gases in a comet. If all the gas in Halley's comet at any instant were mixed into the earth's atmosphere, it would add only 1 atom for each 10,000,000,000 atoms in the atmosphere. Besides, molecules in the coma gases would be stopped by molecular collisions with terrestrial atmospheric gases at a height of 150 to 200 km above the earth's surface. Only a small fraction of the coma's atoms are dangerous, anyway. Unfortunately, the whole scare was started by well-meaning scientists who overstated the situation.

The 1910 appearance of Halley's comet was a time of a great many interesting phenomena. Can we expect as much from the 1985–1986 appearance? The answer is, probably not. Will the comet be a spectacular sight? Again, the answer may be no. To throw some light on these answers, let's look at the case of Comet Kohoutek.

THE KOHOUTEK INCIDENT

In March, 1973, the Czech astronomer Luboš Kohoutek was carrying out a study of asteroids using the 32-inch Schmidt camera at the Hamburg Observatory in Germany. He was photographing the region of the sky opposite the sun. Any asteroids in that region of the sky will be at their brightest, because that is the direction in which the asteroid belt is closest to the earth at any time. On the night of March 7, Kohoutek noticed a suspicious faint, fuzzy blob of light on one of his plates (Figure 2–2). Two nights later, Dr. Kohoutek once again photographed the same region of the sky, and the fuzzy blob was present, but it had moved a small amount. Kohoutek had discovered Comet 1973f, his second discovery in just eight days! The earlier Comet Kohoutek (1973e) remained very faint, and passed from view in a few weeks. However, the second Comet Kohoutek (1973f) was a different story (see Figure 6–2). After enough observations were obtained, astronomers were able to calculate preliminary orbital parameters. The results indicated that the comet would pass close enough to the earth to be a spectacular sight.

Just how bright would it be? The process of estimating future brightnesses of comets is far less precise than calculating their positions on the sky. If the comet were a solid body with a known size and reflectivity, the problem would be simple. The object's known distance from the sun could be used to calculate how much sunlight would reach it; the size and reflectivity could be used to calculate how much of that light would be reflected back into space; and the known distance from the object to the earth could be used to calculate how much of the reflected light would reach us. The problem lies in the size and "reflectivity." Part of the light from a comet that reaches us is sunlight reflected from the dust and, as discussed in Chapter 6, part is fluorescence from the molecules in the comet. The major question is, How much material is present to reflect sunlight and to fluoresce? The answer to that question depends on the rate of sublimation of the nuclear material as the comet approaches the sun and on the rate that the resulting gas and dust leaves the comet and spreads out into space, rates that are highly uncertain. The upshot of this discussion can be stated simply: estimates of future cometary brightnesses are highly uncertain.

One of the earliest estimates of the brightness of Comet Kohoutek when it would be at its brightest in December, 1973, was extremely optimistic: the comet might be bright enough to be seen in broad daylight. The comet was potentially exciting for other reasons, too. It was clearly a long-period comet, with a period of millions of years. Perhaps it just came in from the Oort cloud. Perhaps it was pristine material from the origin of the solar system. If so, astronomers had the opportunity for a detailed look at such material.

Long-period comets are usually discovered a matter of weeks before they reach maximum brightness. Kohoutek was discovered months in advance, affording scientists plenty of time to plan their study strategy. NASA included the comet in its scientific programs, especially Skylab. The media picked up on it, and excitement was raised in all segments of the population. As more data came in, it became clear that the comet would not be as bright as initially estimated. The "Christmas comet" turned out to be barely visible and was a great disappointment to the public.

By the spring of 1974, the event was over. There were occasional articles in the papers about the "Christmas dud." Why? Kohoutek was a scientific spectacular. Several new molecules (CH_3CN, HCN) were identified for the first time. Studies of the comet's coma and tail taught us a number of interesting new facts about their structure, particularly the dynamics of the tail.

The reason the comet was considered a dud by the media is best illustrated by our personal experiences. One of us (J.C.B.) spent January, 1974, on a New Mexico mountaintop observing the comet; the other (R.D.C) remained in the Washington, D.C. suburbs. Brandt saw the comet as a spectacular object on several nights and as a naked eye object nearly every night for a month. Chapman virtually never saw it. Air pollution, city lights, and northeastern weather were too much for the comet. Unfortunately, most of the population lives in urban or suburban settings where pollution has a strong, adverse effect on viewing the heavens.

As we look at the promise of Halley's comet for the 1985–1986 return, we should keep the Kohoutek incident in mind.

COMET HALLEY, 1985–1986

The first look at Halley on its 1985–1986 return came in mid-October, 1982 (Figure 10–6). Working with the great 200-inch telescope on Palomar Mountain, astronomers G. E. Danielson and D. C. Jewett (Figure 10–7) obtained an electronic photograph of the comet in the direction of the constellation Canis Minor as it moved in toward the sun. The comet appeared as little more than a faint (magnitude 24.5!) star-like image to the highly sensitive electronic camera attached to the

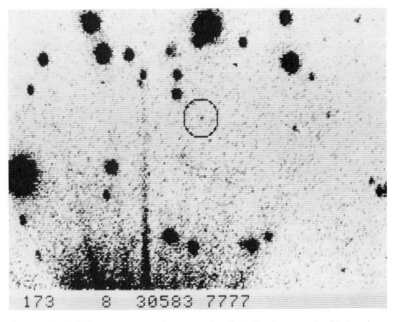

173   8   30583 7777

Figure 10–6. The recovery photograph of Halley's comet obtained on October 16, 1982, using an advanced electronic detector system and the 200-inch Hale telescope at Palomar Observatory (California Institute of Technology photograph).

Figure 10–7. Halley's recoverers, David C. Jewitt (left) and G. Edward Danielson, examine the image obtained on October 16, 1982 (California Institute of Technology photograph).

telescope. It gave itself away as a comet, however: the astronomers obtained a series of images over several nights, and the comet's motion was clearly discernible. When Halley was recovered it was beyond the orbit of Saturn, over a billion miles from the sun. On its 1910 return, the comet was first picked up seven months before perihelion. This time the comet has been recovered more than three years before perihelion passage, a tribute to the great advances in astronomical instrumentation since 1909.

What will the remainder of the return of Halley be like? The comet will pass perihelion on February 9, 1986—a fact that was confirmed by the first few days of study after recovery—at which time it will be 88,500,000 km from the sun and 231,000,000 km from the earth. Unfortunately, it will be on the far side of the sun and will be unobservable at perihelion. Interestingly, the comet will pass close to the earth twice: once on its way in to perihelion and once on its way out from perihelion. The closest approach on the inbound leg of its journey will occur on November 27, 1985, when it will be 93,000,000 km from the earth, and on the outbound leg the closest approach will be on April 11, 1986, when the distance will be 47,900,000 km. Compare these numbers to the closest approach of May, 1910, when the earth passed through or very near the comet's tail and came within 24,000,000 km of the head. Figure 10–8 gives summary diagrams showing the geometrical circumstances of the apparitions of 1759, 1835, 1910, and 1986.

The comet will be visible for several weeks around each of the closest approaches. At the November, 1985, closest approach, the comet will be "above" the ecliptic and will be visible in northern skies. At the April, 1986, closest approach, it will be "below" the ecliptic and will be best seen in the southern hemisphere. Since the April approach is the closer of the two, our southern friends will get the best view of the comet.

How bright will the comet be? The historical record in Halley is long, and as a result we can make some realistic estimates of its most probable brightness as it passes through the inner solar system. Unfortunately, it appears that Halley's comet will not be as bright as Comet Kohoutek.

For those of us who have been waiting a lifetime to see this comet, some planning is in order. First, it is essential to choose an observing site as far away as possible from any population center. The pollution of city lights and industrial smoke will overwhelm the faint comet. Second, it is essential to be as far south as possible in April, 1986 (in November, 1985, the southwestern U.S. will be a good observing site). It would be best to be at or south of the equator. Third, it is important to choose a site that has a high probability of good weather. And last, a locale with an agreeable political situation would be nice. How about American Samoa? Any effort made to see the comet, even if only a trip to the mountains in November or December, 1985, will be rewarding.

The best way to observe the comet visually will be to use binoculars, a small telescope, or even just the naked eye. There are two things to bear in mind. First, make sure that your eyes are well adapted to the dark. Go out into the darkness at least ten minutes, and preferably twenty minutes, before you begin your observations. Second, because the comet will be a relatively large, faint object, you should *not* plan to observe it with a high magnification telescope. With high magnification, the comet will probably fill the entire field of view and there will be little or no contrasting background. The result is that you will have a difficult time seeing anything. In fact, you might be observing the comet without even realizing it because of the lack of contrast. A pair of 7 × 50 binoculars will be ideal for observing the comet. The combination of a 50-mm diameter objective lens and a 7-power eyepiece is optimum for the viewing of faint objects. A 7 × 35 binocular or other type would also be quite useful. When the comet is nearest and brightest, it may be larger than the field of view of the binocular

Figure 10–8. The orbit of Comet Halley on its last four passes near the sun and the earth. The diagrams are known as bipolar plots: they show the motion of the comet in a coordinate system that is rotating at the same rate that the earth orbits around the sun. Therefore, the line from the sun to the earth (marked by a large dot) is shown in a fixed position. The comet appears to spiral around the sun as a result of the earth's orbital motion. The earth's orbit is shown by the ⊕ symbol. In each diagram, the path through space followed by Comet Halley is shown as a heavy line. The trace of that path projected onto the plane of the earth's orbit is shown by a dotted line. The direction of the comet's tail is indicated by small comet drawings. Perpendiculars from the comet to the plane of the earth's orbit are shown for selected positions (portions of perpendiculars below the earth's orbit are dashed). A circle 3 or 5 A.U. in diameter in the plane of the earth's orbit is plotted on several parts of the figure for clarity.

In diagrams (a) through (d), point 1 represents the position of the comet when it was recovered. Note how this point has moved farther from the earth with each subsequent pass. This difference shows how astronomical instrumentation and techniques have improved over roughly two and one-quarter centuries (M.B. Niedner, Laboratory for Astronomy and Solar Physics, NASA/Goddard Space Flight Center, Greenbelt, Maryland).

(a, opposite) The path of Comet Halley, 1758 and 1759, as seen from a point 15° above the plane of the earth's orbit. The comet was recovered on December 25, 1758, when it was relatively close to the earth. The comet was above the plane of the earth's orbit when it was recovered and dipped below the plane between April 9 and April 24, 1759. The dates for the numbered positions are:

1. Dec. 25, 1758	*6. Mar. 10, 1759*	*10. May 9, 1759*
2. Jan. 9, 1759	*7. Mar. 25, 1759*	*11. May 24, 1759*
3. Jan. 24, 1759	*8. Apr. 9, 1759*	*12. June 8, 1759*
4. Feb. 8, 1759	*9. Apr. 24, 1759*	*13. June 23, 1759*
5. Feb. 23, 1759		

(b, opposite) The path of Comet Halley during 1835 and 1836, as seen from a point 15° above the plane of the earth's orbit. The dates for the numbered positions are:

1. Aug. 5, 1835 (recovery)	*8. Nov. 18, 1835 (approximate perihelion)*	*14. Feb. 16, 1836*
2. Aug. 20, 1835		*15. Mar. 2, 1836*
3. Sep. 4, 1835	*9. Dec. 3, 1835*	*16. Mar. 17, 1836*
4. Sep. 19, 1835	*10. Dec. 18, 1835*	*17. Apr. 1, 1836*
5. Oct. 4, 1835	*11. Jan. 2, 1836*	*18. Apr. 16, 1836*
6. Oct. 19, 1835	*12. Jan. 17, 1836*	*19. May 1, 1836*
7. Nov. 3, 1835	*13. Feb. 1, 1836*	*20. May 16, 1836*

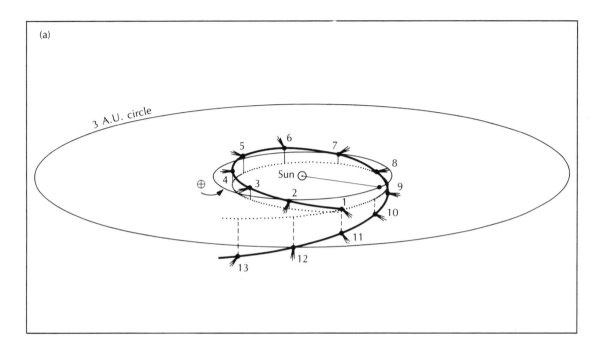

(a)

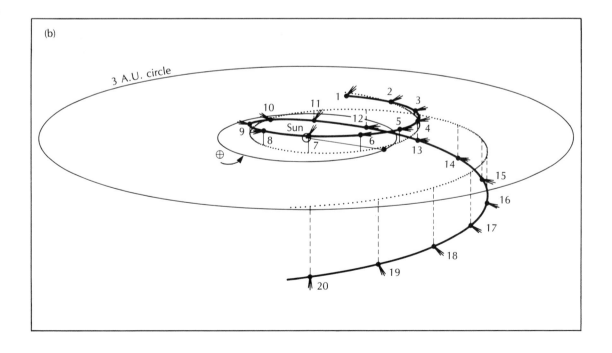

(b)

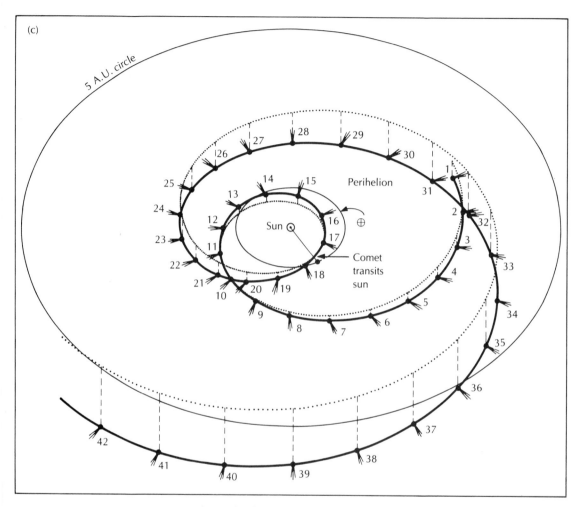

(c) The path of Comet Halley during 1909, 1910, and 1911, as seen from a point 45° above the plane of the earth's orbit. The dates for some of the numbered positions are:

1. Sep. 11, 1909	15. Apr. 10, 1910	29. Nov. 6, 1910
2. Sep. 27, 1909	16. Apr. 25, 1910	31. Dec. 6, 1910
4. Oct. 27, 1909	17. May 10, 1910	33. Jan. 5, 1911
6. Nov. 26, 1909	18. May 25, 1910	35. Feb. 4, 1911
8. Dec. 26, 1909	19. June 9, 1910	37. Mar. 6, 1911
9. Jan. 10, 1910	21. July 9, 1910	39. Apr. 5, 1911
11. Feb. 9, 1910	23. Aug. 8, 1910	41. May 5, 1911
13. Mar. 11, 1910	25. Sep. 7, 1910	
14. Mar. 26, 1910	27. Oct. 7, 1910	

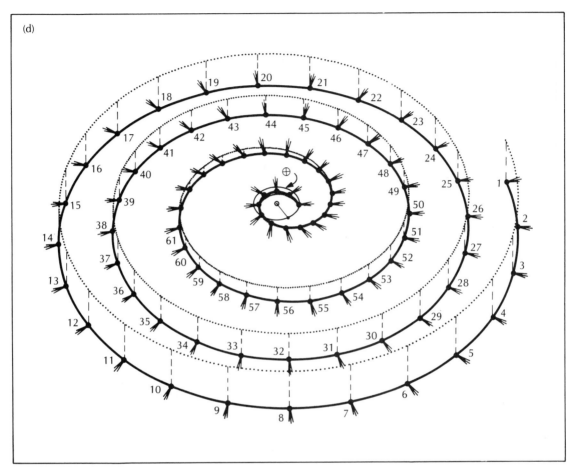

*(d) The path of Comet Halley during 1982–1986, as seen from a point 45°
above the plane of the earth's orbit. The dates for some of the numbered
positions are:*

1. Oct. 16, 1982	22. Aug. 27, 1983	43. July 7, 1984
4. Nov. 3, 1982	25. Oct. 11, 1983	46. Aug. 24, 1984
7. Jan. 14, 1983	28. Nov. 25, 1983	49. Oct. 5, 1984
10. Feb. 28, 1983	31. Jan. 9, 1984	52. Nov. 19, 1984
13. Apr. 14, 1983	34. Feb. 23, 1984	55. Jan. 3, 1985
16. May 29, 1983	37. Apr. 8, 1984	58. Feb. 17, 1985
19. July 13, 1983	40. May 23, 1984	61. Apr. 3, 1985

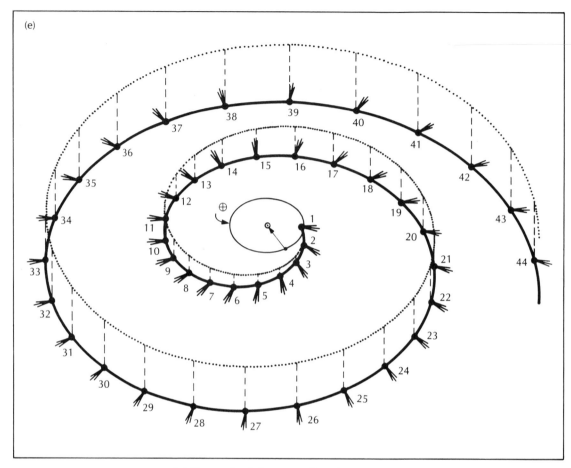

(e) The path of Comet Halley during 1986–1987, as seen from a point 45°
above the plane of the earth's orbit. The last point of part (d) and the first
point of part (e) is March 14, 1986, the nominal Giotto encounter time.
The dates for some of the numbered positions are:

5. May 14, 1986	20. Dec. 25, 1986	35. Aug. 7, 1987
8. June 28, 1986	23. Feb. 8, 1987	38. Sep. 21, 1987
11. Aug. 12, 1986	26. Mar. 25, 1987	41. Nov. 5, 1987
14. Sep. 26, 1986	29. May 9, 1987	44. Dec. 20, 1987
17. Nov. 10, 1986	32. June 23, 1987	

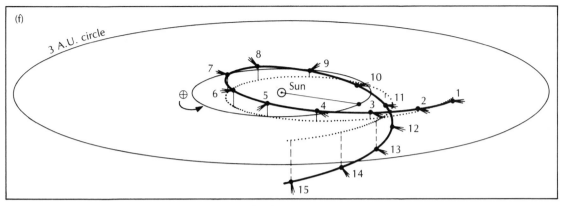

(f) The inner part of the path of Comet Halley during 1985 and 1986, as seen from a point 15° above the plane of the earth's orbit. The dates for the numbered positions are:

1. Nov. 1, 1985
2. Nov. 16, 1985
3. Dec. 1, 1985
   (approximate closest approach to earth)
4. Dec. 16, 1985
5. Dec. 31, 1985
6. Jan. 15, 1986
7. Jan. 30, 1986
8. Feb. 14, 1986
   (approximate perihelion
9. Mar. 1, 1986
10. Mar. 16, 1986
    (approximate fly-by week)
11. Mar. 31, 1986
12. Apr. 15, 1986
    (approximate closest approach to earth)
13. Apr. 30, 1986
14. May 15, 1986
15. May 30, 1986

and you may have to move the binocular around to see the entire comet. If you have a wide-angle telescope, use the lowest magnification that you have. This is one case where low magnification is best.

The chart in Figure 10–9 shows the portion of the sky where the comet will be observable. Its path among the constellations and the dates for several positions are indicated. The next five figures show the position of the comet in the night sky at the end of astronomical twilight in the evening or at the beginning of astronomical twilight in the morning. At those times, the sun is 15° below the horizon and the sky is dark. The horizontal axis in each figure is the terrestrial horizon, with azimuth indicated. Azimuth is measured in degrees from the north point around the horizon toward the east. An azimuth of 90° is due east, 180° is south, 135° is southeast, and so on. Figures 10–10 through 10–12 show the position of the comet in early 1986 for an observer in the northern hemisphere at roughly the latitude of San Francisco, St. Louis, or Washington, D.C. The size of the comet drawing is indicative of the size it is expected to be on the sky. The number in parentheses after each date is its expected magnitude. Remember

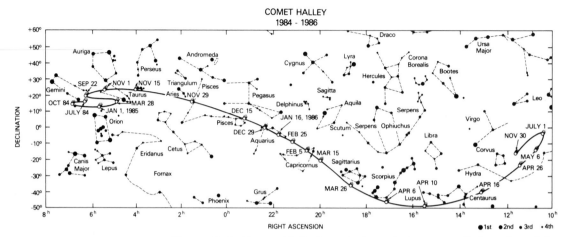

Figure 10–9. Comet Halley's path through the constellations from July, 1984 through November 30, 1986 (Laboratory for Astronomy and Solar Physics, NASA/Goddard Space Flight Center, Greenbelt, Maryland).

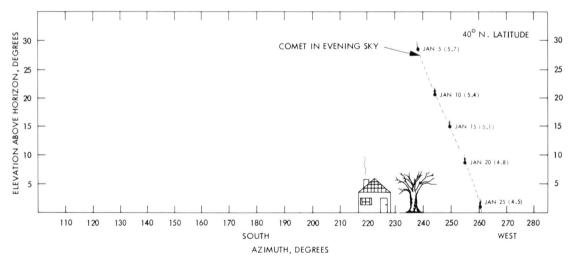

Figure 10–10. Comet Halley's position during January, 1986 for a northern hemisphere observer. The time is at the end of astronomical twilight, so that the sky will be very dark if you are away from city lights and air pollution. (Courtesy of the International Halley Watch, NASA/Jet Propulsion Laboratory, Pasadena, California.)

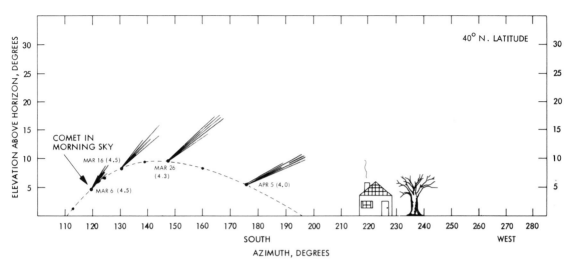

Figure 10–11. Comet Halley's position during March and early April, 1986 for a northern hemisphere observer. The time is the beginning of astronomical twilight. The sky will begin to lighten as the sun rises. This represents the moment that is most ideal for observing. This is also the period when the comet will be at its largest and brightest. In the time since January (Figure 10–10), the comet has passed behind the sun. (Courtesy of the International Halley Watch, NASA/Jet Propulsion Laboratory, Pasadena, California.)

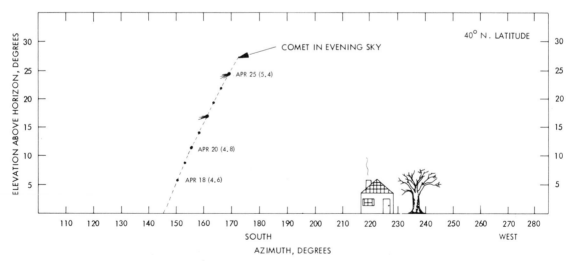

Figure 10–12. Comet Halley's position during April, 1986 for a northern hemisphere observer. The time is the end of astronomical twilight. (Courtesy of the International Halley Watch, NASA/Jet Propulsion Laboratory, Pasadena, California.)

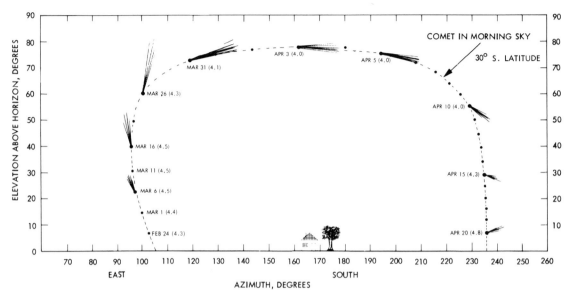

Figure 10–13. Comet Halley's position from late February through mid-April, 1986 for a southern hemisphere observer. The time is the beginning of morning twilight, the time when the sky will begin to brighten. Unlike the case for northern observers, the comet will be high in the sky when it is at its brightest, and will be observable quite a few hours before twilight. (Courtesy of the International Halley Watch, NASA/Jet Propulsion Laboratory, Pasadena, California.)

that the brightest stars are magnitude 1 and the faintest stars visible to the unaided eye are magnitude 6. Remember, also, that the magnitude refers to the total brightness of the comet. A first magnitude star is very bright as a star. However, if its light is smeared out over an area the size of the comet it will seem much fainter. Note that when the comet is at its brightest it will be only 10° above the horizon, at the beginning of morning twilight.

Figures 10–13 and 10–14 show the position of the comet for a southern hemisphere observer at roughly the latitude of Brisbane, Australia; Pretoria, South Africa; or Santiago, Chile. When the comet is at its brightest it will be nearly overhead in these areas.

The 1985–1986 appearance of Halley's comet is not favorable for general viewing, but it will be an important event for the scientific community. As we will see in the next chapter, a great deal of planning has already gone into proposed research efforts.

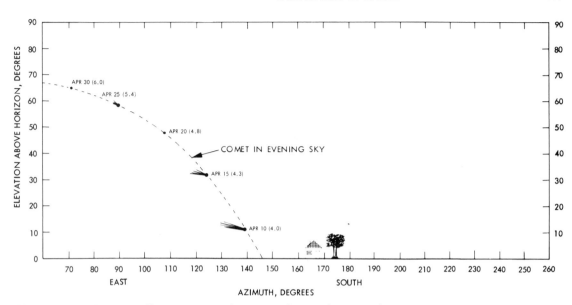

Figure 10–14. Comet Halley's position during April, 1986 for a southern hemisphere observer. The time is just at the end of astronomical twilight in the evening. (Courtesy of the International Halley Watch, NASA/Jet Propulsion Laboratory, Pasadena, California.)

PHOTOGRAPHING A COMET

You have now read a great deal about cometary phenomena that you can try to observe for yourself. You might also find it rewarding to try to photograph a comet, using a fast black-and-white film or one of the high-speed color films. When the comet is faint, it may require several minutes to expose a picture properly. If the camera is set up on a fixed tripod, the rotation of the earth will smear the comet and the stars in a few minutes. Plate 1, showing Comet West, was obtained in this manner. In fact, the over-all composition, with the orange glow of twilight, is so pleasing that the smearing is a minor problem.

It is possible to compensate for the earth's rotation by using a telescope with a clock drive. In general, when a comet is nearby, the wide angle afforded by a single lens reflex with a standard 50-mm lens may be needed to capture the entire comet. Thus we do not necessarily advocate photographing through the telescope—merely attach the camera to the telescope and use the telescope as a moving tripod to track the comet. You will note that some of the pictures of comets in this book (for example, Figure 7–10) show a comet very clearly but the stars are slightly trailed. This is a sophisticated process where the astronomer moves the telescope slightly but constantly to track the comet, while the telescope itself is set to track the stars. This would be a difficult process with most telescopes available to amateur astronomers.

Exposure times are difficult to estimate, and we strongly recommend a series of exposures at say $\frac{1}{16}$, $\frac{1}{4}$, 1, 4, and 16 times the estimated exposure. Experience of acquaintances who have taken exposures of bright comets using a wide-open standard lens such as a single lens reflex and Tri-X film show that a moderately bright, naked eye comet can be recorded in about 15 seconds. An exposure of this length would be an appropriate starting point when Halley reaches naked eye brightness early in 1986.

CHAPTER 11

A VISIT TO A COMET

As the earlier chapters of this book show, cometary science has been active for a long time and much cometary knowledge has been accumulated. However, there are still many unanswered questions. No close-up photograph of a comet nucleus has ever been obtained. The principal parent molecules, besides water, are not known with certainty. The nature of the interaction of the cometary gas with the solar wind is poorly understood. The mechanism by which atoms and molecules in the cometary plasma are stripped of electrons is unknown, as is the magnitude of the magnetic field captured by the comet from the solar wind. Many areas of cometary physics would benefit tremendously from a spacecraft sent to the immediate vicinity of a large, active comet to obtain resolved photographs of the nuclear region and *in situ* measurements of the comet's gas, dust, and magnetic field.

Contemporary space investigations can be wondrous, but they require considerable advance planning and they are expensive. With these practical considerations in mind and considering the desired cometary science, what are the broad criteria for choosing a comet to be visited by a spacecraft?

1. The comet should be bright. The phenomena of the comet should be easily visible from the spacecraft and from the earth. Observations from the earth are most important because the new information from the close encounter needs to be integrated into the pre-existing body of cometary knowledge.

2. The comet should exhibit all known cometary phenomena. The direct mission is very expensive and requires years of planning and effort. It should not be wasted, at least for the first mission, on a mediocre comet.

3. The comet should be famous, a household word. The large expense involved in direct missions means that a major sales effort is

necessary. Public recognition of the comet and public support would assist the sales process.

4. It must be physically possible to send the spacecraft to the comet. On the one hand, this means that launch vehicles must be capable of sending a spacecraft to meet the comet in its orbit. On the other hand, we must know well in advance when to launch the spacecraft, so the comet's orbit must be well known and predictable.

These criteria have an interesting and unique application: Halley's comet. It is bright. A perihelion distance of 0.6 A.U. guarantees a bright comet and generally good viewing circumstances from the earth. Halley's comet exhibits all known cometary phenomena and has a well-known, predictable orbit that can be reached by spacecraft. Not only is it famous, it may be the only universally famous comet. Astronomer Brian Marsden has been quoted as stating the first principle of astrosociology as, "To the man in the street, the solar system consists of Mars, the rings of Saturn, and Halley's comet." Missions have been sent to Mars and to the rings of Saturn and, hence, surely the time for Halley's comet has come. As we have seen, the comet will pass through the inner solar system during late 1985 and early 1986, with perihelion on February 9, 1986.

THE UNITED STATES EFFORT

Many of the frustrations involved in the United States effort to send a spacecraft to the vicinity of Halley's comet are summed up by comet physicist Paul Feldman in a review article: ". . . planning for the now nonexistent NASA mission to Halley was begun as early as 1970!" (*Science* 219:247 (1983).)

The planned missions to comets in the early 1970s were fairly modest. In the latter half of the 1970s and terminating in 1981, the effort focused on more ambitious missions. There were three of them. The first was a Halley Rendezvous Mission, which was to consist of a spacecraft powered by a low, continuous thrust propulsion system called ion drive. The spacecraft would have been sent to the comet on an orbit that could be adjusted to match the comet's orbit once it reached the comet. The spacecraft could then maneuver in the vicinity of the comet for months, carrying out measurements and observations at different distances from the nucleus. Measurements would be made of the gas (composition, density, outflow speeds, and temperature), dust (amount and composition), and magnetic field (strength and direction). Observations would concentrate on imaging of the nucleus to determine its size, shape, and localized structure and to see how these changed with time during the period of the rendezvous. We know that comets are very dynamic bodies and clearly observable changes were expected. Plans for the Halley Rendezvous Mission collapsed because of the time and expense needed to develop a new propulsion system. This requirement was dictated (in part) by the fact that the orbit of Halley's comet is retrograde, that is, basically it revolves around the sun in a direction opposite to the planets. In simple terms, for a spacecraft to rendezvous with the comet after being launched from the earth, it must stop going in the so-called normal direction, turn around, and obtain orbit speed in the opposite direction. Besides requiring a new propulsion system, this "turn around" maneuver required time to perform. Two new propulsion systems were studied— the solar sail, a one-kilometer square sheet of light plastic, and the ion drive; the latter was preferred. When approval to complete the engineering development of ion drive on an urgent basis was not forthcoming, the Halley Rendezvous Mission was no longer possible. It would have to have been launched in May or June, 1982.

Because ion drive was believed to be a very useful propulsion source for many other facets of solar system exploration, there was hope that a mission to Halley's comet with a later launch date might be possible. While studying the options, the celestial mechanicians discovered a marvelous alternative. With ion drive, a spacecraft could be sent to "fly by" Halley's comet and then proceed to a later rendezvous with Comet Tempel 2. This Halley/Tempel 2 Mission had several interesting advantages, even though it was an alternative to the prime candidate, the Halley Rendezvous Mission. The Halley/Tempel 2 Mission became the International Comet Mission (ICM) when a collaboration with the European Space Agency (ESA) became a reality. As part of the collaboration, ESA would build a direct entry probe that would be dropped off and penetrate Halley's coma while the main spacecraft proceeded to Comet Tempel 2. The necessary slip in launch could now be tolerated and launch would occur in June of 1985 instead of May or June of 1982. The ICM would perform the Halley fly-by in November 1985; rendezvous with Tempel 2 would be from July, 1988, to July, 1989. Among the advantages of the ICM was the fact that two comets would be explored in one mission. The American part of this mission never really saw the light of day. We return to the European part later in this chapter.

When the demise of the ICM seemed inevitable, one last attempt was made within NASA to obtain approval for a simpler, less expensive mission to Halley's comet. The Halley Intercept Mission was the result. This mission did not require ion drive and featured an excellent imaging system generally considered superior to others actually going to Halley. High-quality images of the nucleus would be obtained by a narrow-angle camera, while a wide-angle camera would obtain an extremely valuable, almost continuous record of the large-scale state of the comet during an extended "observatory phase." (This latter objective may still be achievable on a limited scale; see below.) The death of the Halley Intercept Mission during the summer of 1981 obtained considerable notoriety in the press (Figure 11–1).

The final attempt at a U.S. mission was a private effort by the group of space enthusiasts who had helped keep the Viking Lander operating on Mars through the "Viking Fund." They formed an analogous "Halley Fund" (Figure 11–2) and hoped to fund a mission through public and private contributions. The challenge was large, the effort was noble, and no one should be surprised that they did not succeed. The money collected will be used to support U.S. cometary science as noted in the ad in Figure 11–2.

MISSIONS TO HALLEY'S COMET

We are delighted to report that other nations have risen to the challenge of direct exploration of Halley's comet during the 1985–1986 apparition and, therefore, we will not have to wait until 2061 A.D. for this process to begin. Three missions, consisting of four or five spacecraft, will encounter Halley's comet in March, 1986. The similarity of the encounter dates results from considerations of celestial mechanics. In this section, we discuss these missions in order of increasing miss-distance from the nucleus.

The European Space Agency's Giotto Mission (named after the painter, mentioned in Chapter 3) has its origins in the entry probe that was to have been ESA's part of the International Comet Mission. The spacecraft (Figure 11–3) will be spinning, and imaging of the nucleus will be accomplished through use of a spin-scan camera. Several im-

Figure 11–1 (opposite). Page 80 of the August 3, 1981, issue of Newsweek. (Copyright © 1981 by Newsweek, Inc. All rights reserved. Reprinted by permission.)

Meet Halley's Comet

GEORGE F. WILL

In 1910, the last time Halley's comet came by, an Oklahoma sheriff had to stop some peculiar citizens from sacrificing a virgin to the comet. The comet is coming again in 1986, so Oklahomans should lock up their daughters. And David Stockman should stop sacrificing science on the altar of parsimony.

I shall use my dying breath to whisper praise of Stockman, but he should not have killed NASA's plan to send a satellite to intercept the comet. It would have cost $300 million over five years, 25 cents per person a year, and it should have been an occasion for the Administration to leaven its frugality with a farsighted exception.

Comets, and especially Halley's, have excited superstition far from Oklahoma. The historian Josephus said a comet resembling a sword (Halley's, in A.D. 66) foretold the destruction of Jerusalem (A.D. 70). The visit of Halley's comet in 1066 was thought to have been a portent of the unpleasantness that befell King Harold at Hastings. Shakespeare said: "When beggars die, there are no comets seen; the heavens themselves blaze forth the death of princes." The day Edward VII died (May 6, 1910) Halley's comet was especially vivid (more vivid than that particular prince merited). Mark Twain, born during the comet's 1835 visit, said he would be disappointed if he didn't depart when it came again. He died April 21, 1910, just before the comet's "tail" brushed earth and as (Twain would have loved this) people were selling anti-comet pills to a public panicky about gases in the tail.

In "The Comet Is Coming! The Feverish Legacy of Mr. Halley," Nigel Calder says Halley's comet is, as most comets probably are, "sky pollution," a "dirty snowball that comes tumbling out of the freezer of twilight space." (There are an estimated 100 billion comets in our itsy-bitsy solar system.)

Flu Machines: These cosmic jaywalkers rarely bump into anything because space is even more vacant than Wyoming. (If there were just three bees in America, the air would be more congested with bees than space is with stars.) But there is a constant rain onto earth of meteoric debris, and an occasional "thump." Calder writes: "Early in the morning of 30 January 1908 the driver of the trans-Siberian express heard loud bangs and imagined that his train had exploded . . .

his wide-eyed passengers said they had seen a bright blue ball of fire . . ." A small comet had leveled a 70-mile-long strip of forest.

But some collisions may have been constructive. One theory is that a comet brought to earth the first bacteria or whatever it was that started the ol' ball of life rolling 4 billion years ago (fortunately, before governments demanded environmental-impact statements). Another theory is that comets are "flu machines," bringing viruses to earth. Ask now what caused the fall of Rome and the rise of Christianity. Calder says some theorists argue: "During the period from A.D. 400 to 1400 the earthlings had a particularly nasty time with the clouds of diseases spun off from comets. A

> *Conservatives cannot turn space exploration over to their two loves, federalism or capitalism.*

'disease-filled' millennium . . . forced people to live farther apart and thus to 'uncivilise' themselves; it also . . . moved the Europeans to adopt the 'sombre' religion of Christianity."

But if that is true, Calder asks, why not now? "If a millionth part of the meteoric debris falling to earth from comets consists of viruses, a small garden could collect millions of viruses every day, ready to assail plants, pets and humans."

Calder finds a bit more plausible the theory that a comet killed the dinosaurs; they did die out suddenly, and folks used to think they were just too big for Noah's ark. Today some scientists think a big comet, perhaps 6 miles in diameter, struck earth, throwing up a hundred times its weight in dust—much more dust than was sent up by the eruption in 1883 of the Krakatau volcano, which produced "glorious sunsets" around the world for two years. The theory is that the cloud produced by the comet collision blocked out sunlight, and in the four-year "night" much vegetation and most dinosaurs died.

Why, then, is there no crater? Well, there

is a suspicious ring-shaped something on the seabed north of Australia. (Oceans and continents have been meandering around during the last 65 million years.) And in geological formations around the world are thin layers of clay with a chemical composition that suggests that long ago the earth was suddenly swamped with a particular element (iridium) in an amount that seems unlikely to have come from a source on earth. If this theory is true, then if, 65 million years ago, the comet had come by an hour earlier or later, it would have missed and dinosaurs might still rule the earth. So a comet may have been a benefactor.

Mysteries: Anyway, comets are owed the respect due the elderly. Most comets in our solar system spend most of their time loitering (relatively speaking) beyond the outer planets. So they are among the "oldest," meaning least changed, objects: they experience less of the erosion and evolution that erases the imprint of the birth of the solar system. A rendezvous with one might reveal evidence about the origins of the universe, the human race and Oklahoma.

If our curiosity about such things atrophies, so will our humanity. That is why the Halley's comet intercept program, which can still be saved, should be used by the Administration as an opportunity to practice "creative exceptionalism." The country wants conservatism, but needs conservatism subtle enough to make exceptions to the principle of parsimony. Conservatives cannot turn space exploration over to their two loves, federalism or capitalism—to the states or the private sector. Neither Utah nor Exxon can do it. Only Big Government—only our government—can do it.

Conservatives are supposed to take the long view, and to take intangibles seriously. They should want to look back toward the creation of the universe that has produced, as its crowning glory, the Reagan Administration. And they should look far into the future and imagine a future in which mankind is not curious about the wondrous mysteries of its situation.

We know next to nothing about virtually everything. It is not necessary to know the origin of the universe; it is necessary to want to know. Civilization depends not on any particular knowledge, but on the disposition to crave knowledge.

Figure 11–2. Some public efforts to support NASA's comet program. (Top) An advertisement placed by The Halley Fund. (Bottom) The Halley Fund has evolved into The Halley Campaign, 4635 North Fifteenth Street, Arlington, Virginia 22207. The logo is shown here.

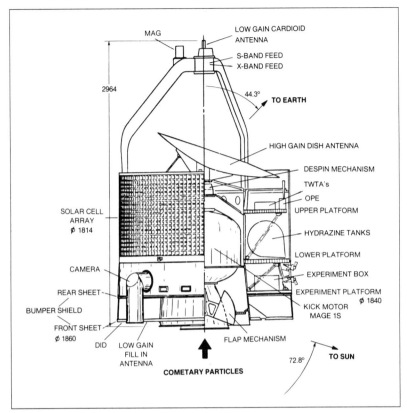

MAG
LOW GAIN CARDIOID ANTENNA
S-BAND FEED
X-BAND FEED
2964
44.3° **TO EARTH**
HIGH GAIN DISH ANTENNA
DESPIN MECHANISM
TWTA's
OPE
UPPER PLATFORM
SOLAR CELL ARRAY
∅ 1814
HYDRAZINE TANKS
LOWER PLATFORM
CAMERA
EXPERIMENT BOX
REAR SHEET
EXPERIMENT PLATFORM
∅ 1840
BUMPER SHIELD
KICK MOTOR MAGE 1S
FRONT SHEET
∅ 1860
DID LOW GAIN FILL IN ANTENNA
FLAP MECHANISM
72.8° **TO SUN**
COMETARY PARTICLES

Figure 11–3. Schematic of the Giotto *spacecraft. (Courtesy of R. Reinhard, European Space Agency, Noordwijk, The Netherlands.)*

ages of the nucleus should be obtained during the time of closest approach. Additional experiments will measure the properties of the neutral and ionized gas, the dust, the magnetic field, and any energetic particles. The data from these experiments must be telemetered back to the earth in real time because at the targeted miss-distance there is a substantial likelihood that the spacecraft will not survive. The trajectory of the *Giotto* spacecraft is illustrated in Figure 11–4.

The encounter distance for *Giotto* is planned to be approximately 1000 kilometers and encounter should occur on March 13, 1986. The geometry of the encounter on this date is such that the comet's distance from the sun is 0.88 A.U.; distance from the earth is 0.99 A.U.; and the sun–earth–comet angle is 53°. Other details of the

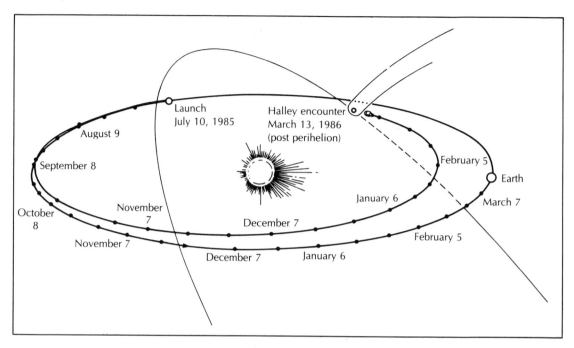

Figure 11–4. The Giotto *trajectory from launch on July 10, 1985, to encounter with Halley's comet on March 13, 1986. (Courtesy of R. Reinhard, European Space Agency, Noordwijk, The Netherlands.)*

encounter are shown in Figure 11–5, and Figure 11–6 shows an artist's rendition of the encounter. These circumstances are good descriptions of the other encounters (discussed below), because the dates differ from March 13 by at most a few days. All the Halley encounters are planned for the sunward side of the nucleus.

The Soviet Union is planning to send two spacecraft (called *VEGA*) to pass within 10,000 kilometers of Halley's nucleus. Each spacecraft will first swing by the planet Venus and will then reach Halley on March 10/11, 1986 and on March 18, 1986. The instrumentation is extensive and generally similar to *Giotto*'s instrumental complement. The imaging is more ambitious, however, and consists of two cameras reminiscent of the ICM imaging system; no extended "observatory phase" is planned.

The third entry is the *Planet A* Mission (Figure 11–7) under preparation by the Japanese Space Agency. The plan is for the spacecraft to encounter Halley's comet at a distance of approximately 100,000 kilometers on March 8, 1986. The principal experiment is an ultraviolet imaging device designed to obtain images at the resonance

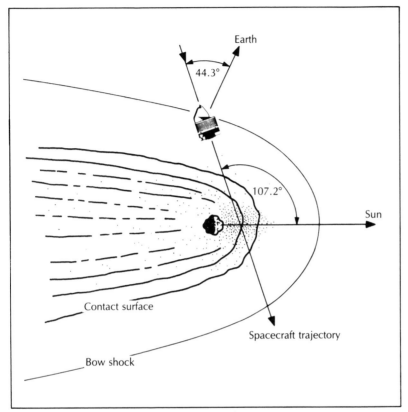

Figure 11–5. The Giotto *encounter geometry. The contact surface and bow shock are physical features caused by the comet's interaction with the solar wind. The contact surface separates the pure cometary plasma (inside) from the mixed cometary and solar wind plasma (outside). The bow shock is produced when a supersonic flow encounters an obstacle; physically analogous features are the bow waves produced by ships and the bow shock seen in photographs of projectiles. (See Figure 11–6 for an artist's rendition of the* Giotto *spacecraft approaching the comet.) (Courtesy of R. Reinhard, European Space Agency, Noordwijk, The Netherlands.)*

line of neutral atomic hydrogen (Lyman-α, 1216 Å). In addition, some plasma measurements will be made. A test spacecraft for *Plant A* may be the fifth sent to the vicinity of Comet Halley.

The direct exploration missions described here are very powerful, particularly when taken together, and hold the promise of great progress in our understanding of comets. A major facet of the activities described in the next section is the desire and intent to enhance and support the results from these pioneering missions.

Figure 11–6. Artist's rendering of the Giotto *spacecraft a few hours before the encounter with Halley's comet. (Courtesy of R. Reinhard, European Space Agency, Noordwijk, The Netherlands.)*

OTHER HALLEY ACTIVITIES

The United States' contribution to the Halley effort consists primarily of ground-based and earth-orbiting efforts. First, we have the International Halley Watch (IHW), an organization with headquarters at the NASA/Jet Propulsion Laboratory, Pasadena, California (Figure 11–8). A headquarters for the eastern hemisphere is located at the Remeis Observatory, Bamberg, West Germany. The IHW has been named by

Figure 11–7. Protomodel of the spacecraft Planet A. *The small horizontal cylinder mounted on the left side of the top plate is the mirror housing of the ultraviolet imaging device for obtaining an image of Halley's hydrogen cloud. The two parallel slots at the right-hand edge of the main body are the inlets for the solar wind particle detectors. The high-gain antenna on top is mechanically despun. The flight spacecraft will have solar cells around the lower part of the body. (Courtesy of K. Hirao, Institute of Space and Astronautical Science, Tokyo, Japan.)*

Figure 11–8. The logo of the International Halley Watch, IHW.

the International Astronomical Union as the official coordinating organization for this apparition. The task of the IHW is to act as a coordinating body for many of the Halley activities described here and in the preceding section, and to organize ground-based observing networks in the major observing techniques in order to provide continuous coverage of the comet during selected intervals of time. The networks are charged with obtaining wide-field imagery (for large-scale phenomena), spectroscopy, photometry, infrared observations, radio observations, narrow-field imagery (for near-nucleus phenomena), and astrometry. One of the authors (J.C.B.) is Discipline Specialist for the Large-Scale Phenomena network. Approximately 90 collaborating observatories around the globe in both the northern and southern hemispheres are required to obtain the desired continuous coverage. The astrometry network serves as an example of the collaboration between the IHW and the other investigations. The astrometry network will be the most likely source of positional data on Halley's comet for the missions to it.

Another major effort involves the Spacelab missions. The OSS-3 instrument complement had been previously selected to carry out astronomical observations in the ultraviolet. The instruments consist of an imaging device, a spectrograph, and a spectropolarimeter. NASA convened a "Science Working Group for Spacelab Observations of Comet Halley." This Working Group noted that the OSS-3 instrument complement, now renamed Astro 1, could make an excellent contribution to cometary science (Figure 11–9). In addition, the Working Group recommended to NASA that a modest, wide-field imaging facility be added for the comet missions. Such an imaging facility would record the large-scale state of the comet with a time between exposures of approximately two hours or less, and greatly assist in the interpretation of the ultraviolet observations made from Spacelab and in the interpretation of the *in situ* measurements made by the direct comet missions. In addition, a high time-resolution record obtained over a period of days would be unique in the history of comet observations and would provide a record, most valuable in its own right, suitable for probing the large-scale evolution and morphology of Halley's comet as well as the astrophysical plasma processes involved. The highest priority would be to cover the encounter date of ESA's *Giotto* Mission, March 13, 1986. A mission of approximately one week's duration would permit coverage of the *Planet A* Mission on March 8 and the first *VEGA* spacecraft on March 10 or 11. Thus, the ASTRO 1 mission should supply important coverage from earth orbit for most of the missions to Halley's comet.

A NASA entry into *in situ* measurements is *ISEE*-3 (*International*

Figure 11–9. Artist's rendering of the Astro 1 payload, on board the space shuttle, observing Halley's comet in 1986. (By now you should know enough about comets to suspect that the artist didn't know which way comet tails point!) (National Aeronautics and Space Administration, Washington, D.C.)

Sun-Earth Explorer). This spacecraft was in orbit near the earth making measurements of the solar-wind plasma, the solar-wind magnetic field, and high-energy particles. It was diverted on December 22, 1983 by a pass only 116 km above the lunar surface. The spacecraft was renamed the *International Cometary Explorer (ICE)*. Its present orbit passes through the tail of Comet Giacobini–Zinner on September 11, 1985. This target is an active short-period comet with a narrow ion tail some 10^6 km in length (Figure 11–10). The plasma and magnetic field experiments on *ICE* could provide valuable data on the ion tail and the solar wind–comet interaction. In addition, after the encounter with Giacobini–Zinner there are two times when the spacecraft will be upstream of Halley and close to the sun–Halley line: October 31, 1985 and March 31, 1986. On these dates, the spacecraft will record the properties of the solar wind that will reach Halley's comet a short time later.

Besides the major efforts just described, there are other observations that could be made by earth-orbiting and other spacecraft that

Figure 11–10. Comet Giacobini–Zinner near perihelion in October of 1959. (Photograph by E. Roemer, University of Arizona, Tucson, Arizona; official U.S. Navy Photograph.)

may be in operation in 1985 and 1986. Perhaps the best known is the *Hubble Space Telescope* currently scheduled for launch in mid-1986. The *Hubble Space Telescope* has the large light-gathering power of a 94.5-inch primary mirror and features an instrument complement of imaging cameras, spectrographs, and a photometer. Major contributions to the study of Halley's comet can be expected from the *Hubble Space Telescope* (Figure 11–11). In particular, its powerful instruments should be able to observe Halley's comet throughout the comet's orbit. Thus 1982 could be the last time that Halley's comet is "recovered" in the traditional sense.

Contributions can also be expected from the Solar Maximum Mission, which was repaired in 1984, from the *Galileo* spacecraft (if it

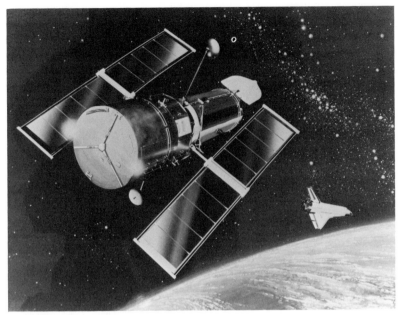

Figure 11–11. Artist's rendering of the Hubble Space Telescope in earth orbit. With this launch, an interesting milestone will be achieved in cometary studies. Instruments on the Hubble Space Telescope are sufficiently sensitive to record Halley's comet all the way around its orbit. (National Aeronautics and Space Administration, Washington, D.C.)

is sent to Jupiter on a suitable trajectory), and from the *International Ultraviolet Explorer* (if it continues to function over seven years after its launch). The important point is that the significance of Halley's comet and the importance of the opportunity presented at the 1985–1986 apparition guarantee that nearly all available facilities will participate in the scientific onslaught on the nature of our best-known comet.

FURTHER READING

If this book piqued your interest in comets, then you might be interested in looking at the following books.

Brandt, J.C. and R.D. Chapman. 1981. *Introduction to Comets*. Cambridge and New York: Cambridge University Press. This monograph is the most up-to-date advanced text about comets on the market. It contains an extensive, annotated bibliography (over 200 entries) of the technical literature. (Available as a paperback.)

Brandt, J.C. 1981. *Comets*. San Francisco: W.H. Freeman and Company. A compendium of relevant articles from *Scientific American*.

Wilkening, L.L. (ed.). 1982. *Comets*. Tucson: The University of Arizona Press. An up-to-date collection of technical reviews of various aspects of cometary science by a number of well-known researchers in the field.

The popular astronomy magazine *Sky and Telescope* runs a regular monthly column that discusses current comet news.

GLOSSARY

absorption line   A narrow, dark gap (lower light intensity) in the continuous spectrum (higher light intensity) of a source. It can arise when the source is a glowing solid seen through a cooler gas, or under more complex circumstances. The presence of specific patterns of absorption lines is diagnostic of the presence of certain atoms, ions, or molecules in the cooler gas.

adsorption   The process in which atoms or molecules of a gas or a liquid adhere to the surface of a solid. The surface includes areas exposed in microscopic cracks and fissures. It differs from absorption in which molecules of one substance mix with the molecules within a solid body.

Ångström   A unit of length equal to 0.00000001 or 10^{-8} centimeters.

anti-tail   A cometary tail that appears to point toward the sun.

aphelion   The point in the orbit of a solar system body (such as a planet or a comet) that is farthest from the sun.

ascending node   The point in the orbit of a comet (or other solar system body) where the body passes from below (south of) the ecliptic to above (north of) the ecliptic.

asteroid   One of many small bodies orbiting the sun primarily between the orbits of Mars and Jupiter. Sizes range from 950 km in diameter (Ceres) to the minimum observed size, a few tenths of a km. Sometimes called "minor planets."

astronomical unit (A.U.)   A unit of length equal to the earth's average distance from the sun. An astronomical unit is approximately 150,000,000 km or 93,000,000 miles.

bow shock   A shock wave caused when the supersonic solar wind encounters a comet as an obstacle; analogous to the shock wave caused by a supersonic aircraft or the wave caused by a boat moving through water.

black body   A hypothetical body that absorbs all of the electromagnetic radiation that falls on it.

celestial mechanics   The science of the motions of celestial bodies.

coma   The roughly spherical halo of neutral gas and dust surrounding the nucleus of a comet.

constellation   One of more than 80 arbitrary groupings of stars, which people imagine to look like an object, an animal, or a mythological person.

contact surface   The surface in the interaction region between a comet and the solar wind. It separates pure cometary plasma on the inside with mixed cometary and solar-wind plasma on the outside.

corona   The superhot, low density outer layer of the sun. It extends from just above the solar surface throughout the inner solar system.

coronagraph   A specially designed solar telescope that is equiped with a device to block out the bright photospheric light of the sun, permitting observation of the much fainter corona. A coronagraph can be effective only at the very best observing sites, where the atmosphere is very clear, or in spacecraft above the atmosphere.

desorption   The opposite of adsorption. The release of molecules adhering to the surface of a solid object.

despun   Any portion on an otherwise spinning spacecraft that is not spinning is said to be despun. The *Orbiting Solar Observatory*, a NASA spacecraft, consisted of a so-called wheel portion that was spinning and a sail portion that was always oriented toward the sun. The sail was despun.

direct motion   The counterclockwise motion of the planets around the sun as seen from above the north pole of the earth.

disconnection event   A spectacular event in the life of a comet when an existing tail is completely detached from the head, probably because the comet interacts with a boundary in space where the interplanetary magnetic field changes direction over a short distance.

Doppler effect   A change in the frequency of sound or light waves caused by relative motion of the source and the observer. For example, the pitch of a siren appears to change as an emergency vehicle rushes past.

dust particle   A small chunk of non-icy solid material. Possible cometary dust particles collected in the atmosphere by high-flying aircraft are irregularly shaped, a few microns in size, and are probably similar in composition to sand.

dust tail   The cometary tail composed primarily of dust. The dust tail has a yellowish color because what is seen is sunlight reflected off small particles of dust.

eccentricity   A parameter used to describe the shape of an orbit. A circular orbit has 0 eccentricity. An elliptical orbit has an eccentricity larger than 0 and smaller than 1. The larger the eccentricity, the more elongated the ellipse appears to be.

ecliptic   The plane of the earth's orbit, or the line on the sky where the earth's orbit intersects the sky. The sun, the moon, and the planets appear to remain close to the ecliptic as they move around the sky.

electromagnetic wave   A wave of radio, infrared, visible, ultraviolet, X ray, or gamma ray radiation.

elliptical orbit   An orbit in the shape of a closed figure that appears oval in shape. Planets, asteroids, and comets have elliptical orbits.

emission line   The narrow line of color in the spectrum emitted by a hot, glowing gas at one or a few selected wavelengths. The specific lines depend on the atom, ion, or molecule present in the gas.

excitation   The process in which electrons bound in an atom gain energy, while remaining bound to the atom. There are a finite number of well-defined excited states in any atom or ion.

false nucleus   An apparent bright spot within the head of a comet. The false nucleus is merely the brightest part of the coma, and not the true nucleus.

first-generation molecule   A molecule that results when sunlight dissociates (tears apart) the parent molecules sublimated from the nuclear ices of a comet.

fluorescence   The process in which atoms, ions, or molecules absorb and then re-emit radiation.

galaxy   A very large celestial object consisting of many billions of stars and vast quantities of gas and dust. The sun is in the Milky Way galaxy.

geocentric parallax   The parallax of an object observed from different locations on the surface of the earth.

Greenwich Mean Time (G.M.T.)   The standard time at Greenwich, England. G.M.T. is used by astronomers all over the world to report their observations so that the times are easily compared. G.M.T. is also called Universal Time (U.T.).

head   The part of a comet consisting of the nucleus, coma, and tail plasma in transit from the nucleus to the tail. To the naked eye, it is the brightest part of the comet.

heliocentric   Centered on the sun. The heliocentric distance of a comet is its distance from the sun.

high-speed stream   A stream of solar-wind plasma that travels with a speed well above the average speed of typical solar wind plasma.

hydrogen–hydroxyl cloud   A giant cloud of gas found by space observations to be surrounding comets. The cloud is composed of atomic hydrogen and the hydroxyl (OH) radical.

hyperbolic orbit   An open orbit. An object in a hyperbolic orbit will never return to a given point in space, but will travel forever in more or less the same direction. See *parabolic orbit.*

inclination   The angle between the orbit plane of a planet, comet, or other celestial object and the plane of the ecliptic.

intermediate-period comet   One of a handful of comets with orbital period between roughly 30 and 200 years.

interplanetary material   The solid particles and solar wind plasma between the planets in the solar system.

interstellar material   The dust and gas between the stars in the Milky Way galaxy.

ion   An atom, molecule, or radical that has a net electrical charge.

ion drive   A rocket engine that uses ions accelerated in an electric field for thrust. An ion drive engine has very low thrust but can operate for years on a small amount of fuel.

isotope   One of several variants of a chemical element that differ only by the number of neutrons in the nucleus. Isotopes all behave identically as far as the chemist is concerned.

long-period comet   A comet with an orbital period longer than 200 years. The orbits of long-period comets are oriented at random in space.

magnitude   A brightness scale used by astronomers. Roughly, the brightest stars in the sky are magnitude 1 and the faintest stars visible to the unaided eye under ideal conditions are magnitude 6. Each step of one magnitude corresponds to a brightness ratio of 2.512. Brighter objects have smaller magnitude numbers.

meteor   A brief streak of light in the sky caused by a small particle burning up in the atmosphere. Also called a shooting star.

meteorite   Part of the object causing a meteor that is large enough to reach the ground.

meteoroid   What the solid object causing a meteor is called while it is still out in space.

meteor shower   Typically, a large number of meteors that seem to originate from a small area of the sky during a brief time period, usually a night or two.

meteor stream   A clump of meteoroids that seem to follow the same orbit in space.

micron   A unit of length equal to one millionth of a meter or 10,000 Ångströms.

minor planet   An asteroid.

miss distance   For our purposes, the distance of closest approach of a space probe to the nucleus of a comet. In general, the distance of closest approach in any encounter situation, such as a *Voyager* spacecraft swinging by Saturn or an electron whizzing past an atomic nucleus.

molecular band   A complex pattern of spectrum lines absorbed or emitted by molecules. The lines are both numerous and closely spaced so that they appear as a broad band of absorption or emission.

molecule   The smallest unit of a chemical compound, composed of a number of atoms bound together. A water molecule is composed of two atoms of hydrogen and one atom of oxygen.

momentum   The quantitative measure of the motion of a body. Momentum is the product of the mass and the velocity of the body.

nongravitational force   A force affecting the motion of a comet. It is due to directional sublimation of the ices of a rotating nucleus.

nucleus   The solid part of a comet, thought to be a kilometer-sized chunk of dusty ice.

Oort cloud   A cloud of comets surrounding the solar system at distances between 50,000 and 150,000 A.U. from the sun.

orbit   The path one body follows around another body. For instance, comets go around the sun in elliptical orbits.

orbital elements   The six parameters that describe the size, shape, and orientation in space of an orbit, and that specify the position of a body in its orbit at one particular instant of time, usually the time of perihelion passage.

osculating orbit   The elliptical orbit a planet or comet would follow if all the perturbations acting on it were switched off at some instant. The osculating ("kissing") orbit just "kisses" the true orbit at that instant. The true orbits of planets and comets depart from exact ellipses because of perturbations.

outburst   A brief period of unusually rapid gas and dust emission from a comet.

outgassing The process in which a material body releases absorbed gas.

parabolic orbit An orbit with an eccentricity of exactly 1. A parabola is the limiting form of an ellipse that has been stretched until its major axis is infinitely long. A comet in a parabolic orbit is moving just fast enough to escape the solar system. By contrast, a comet in a hyperbolic orbit has a velocity that exceeds the minimum needed to escape.

parallax The apparent shifting of a nearby object relative to distant objects when the nearby object is viewed from adjacent positions. If you look at a finger held at arm's length first with one eye and then with the other, the finger will appear to move relative to more distant objects. See also *geocentric parallax*.

parent molecule One of probably many molecules that make up the ices of a cometary nucleus. Parent molecules are quickly torn apart by the ultraviolet light of the sun after they are released from the cometary nucleus.

perihelion The point in the orbit of a comet, planet, asteroid, or other body orbiting the sun that is nearest the sun.

perturbations The small effects of all the other bodies in the universe on the motion of one body orbiting another. The influence of the planets on the motion of a comet are planetary perturbations.

photometry The process of measuring the brightness of light.

photon A "particle" of light with no mass or electric charge. Light sometimes behaves as if it were a stream of such particles.

photosphere The bright surface of the sun.

plasma A gas consisting of an equal number of positive ions and electrons. The behavior of a plasma is different from the behavior of a gas of neutral atoms because of electrostatic interactions between particles in the plasma and possible magnetic effects.

plasma tail The portion of a cometary tail that is composed of plasma. The plasma tail usually has a blue color because of emission bands of CO^+ in the blue part of the spectrum.

polarization A uniform variation of a wave characteristic, as when a light beam oscillates in one plane.

primary mirror The main mirror that gathers light and focuses it into an image in a telescope that uses mirrors as optical components.

protocomet A body in the early solar system that eventually became a comet. A protocomet was probably a large icy body at the distance of the outer planets from the sun.

protoplanet One of many bodies in the early solar system that ultimately formed into the planets.

protosun The early sun before internal nuclear reactions caused it to shine.

radiant A point on the sky from which the meteors in a shower appear to originate.

radiation pressure The tiny force exerted on an object by a beam of light. Radiation pressure is such a small force that it has no effect on bodies of the size we meet in everyday life but it can move small dust particles in space.

radical Part of a molecule. Water is most correctly called HOH, and it can be broken up into a hydrogen atom and the OH or hydroxyl radical.

refractor An astronomical telescope in which the main optical element is a lens; distinct from a reflector in which the main optical element is a mirror.

retrograde loop A small loop in the apparent path on the sky of planets farther than the earth from the sun, caused by a periodic halt, reversal, and then resumption of the planet's slow eastward motion among the stars.

retrograde motion The opposite of direct motion; the apparent clockwise motion of some comets around the sun as seen from above the north pole of the earth.

revolution The motion of a body in its orbit about another body.

rotation The spin of a body on its own axis.

semimajor axis One half of the long axis of an ellipse.

short-period comet A comet with an orbital period shorter than 30 years.

solar sail A gigantic sail (a kilometer on a side)

made of very light material that would use the solar wind to move a spacecraft through the solar system.

solar wind The low-density plasma expanding away from the sun at an average speed of 400 km/sec.

spectrograph A device for breaking or dispersing the light from an object into its constituent colors.

spectroscopy The process of obtaining and studying a spectrum.

spin-scan camera A camera that builds up an image by scanning a small field of view across an object by the rotation of the camera.

spectrum The distribution of energy (light) emitted by a radiant source, arranged in order of wavelengths.

spiral arm A spiral-shaped collection of dust, gas, and stars in a galaxy.

split comet A comet that has broken up into two or more complete comets.

sporadic meteor A meteor that is not part of a shower.

star cluster A closely-associated group of stars bound together by their mutual gravitational interactions.

sublimation The process in which a substance passes from the solid state directly to the gaseous state.

sun-grazing comet One of a group of comets that pass extremely close to the sun at perihelion.

supernova The unimaginably intense explosion of a star at the end of its life, involving a brightness increase of about a billion times.

Swings effect The effect in which the relative strengths of molecular bands in the spectrum of a comet change with the position of the comet in its orbit. The effect is caused by the changing of relative positions of molecular bands in the cometary spectrum and absorption lines in the solar spectrum due to the Doppler effect.

tail The extended portion of a comet consisting of dust or plasma or both and always pointing away from the sun.

tail ray A narrow streak of bright emission in a cometary plasma tail.

transit The process in which a celestial body passes across the disk of the sun.

twilight The period of time before sunrise or after sunset when the night sky is still partially illuminated by sunlight.

type I tail Another name for plasma tail.

type II tail Another name for dust tail.

ultraviolet light Short wavelength light beyond the visible portion of the spectrum. The earth's atmosphere is opaque to most ultraviolet light, and scientists must use satellites sent into orbit to study the ultraviolet emission of celestial objects.

wavelength The distance from crest to crest or trough to trough in a wave.

INDEX

COMET INDEX